Advances in Silicon Science

Volume 5

Series Editor

Janis Matisons
Gelest Inc., 11 East Steel Road
Morrisville, PA 19067, USA

This series presents reviews of the present and future trends in silicon science and will benefit those in chemistry, physics, biomedical engineering, and materials science. It is addressed to all scientists at universities and in industry who wish to keep abreast of advances in the topics covered.

Aims & Scopes

Silicon is unique. Once thought to bridge the gap between organic and inorganic chemistry, it has now gone well beyond such simplistic perceptions. It incorporates reactions that have their own 'organic' chemistry (e.g. hydrosilylation) and covers solid materials that can be either metallic (e.g. silicon metal), semiconductors (e.g. silicon carbide wafers), ceramic (e.g. porous silicon) or inert particulate fillers (e.g. precipitated silica). Silicon is in the same group of the periodic table as carbon, yet silicon carbide is not another form of diamond or graphite or even amorphous carbon black. It, like most other silicon materials, is unique.

This book series represents the journeys of many eminent silicon scientists into the unique world of silicon materials and molecules. As such, it covers the unique silicon chemistry emerging this century; very similar to the position where organic chemistry stood at the emergence of the last century. It explores how silicon structures convey substance to many silicon materials with a molecular precision that defines silicon science and technology.

More information about this series at http: //www.springer.com/series/7926

Paul M. Zelisko
Editor

Bio-Inspired Silicon-Based Materials

Editor
Paul M. Zelisko
Department of Chemistry and Centre for Biotechnology
Brock University
St. Catharines
Ontario
Canada

ISSN 1875-3108 ISSN 1875-3086 (electronic)
ISBN 978-94-024-0534-7 ISBN 978-94-017-9439-8 (eBook)
DOI 10.1007/978-94-017-9439-8
Springer Dordrecht Heidelberg New York London

Softcover reprint of the hardcover 1st edition 2014

Printed on acid-free paper

Springer is part of Springer Science+Business Media (www.springer.com)

Contents

1 **Silicon in a Biological Environment** 1
Paul M. Zelisko

2 **The Role of Silicates in the Synthesis of Sugars Under Prebiotic Conditions** 19
Joseph B. Lambert, Senthil Andavan Guruswamy-Thangavelu

3 **Protease-Mediated Hydrolysis and Condensation of Tetra- and Trialkoxysilanes** 27
Mark B. Frampton and Paul M. Zelisko

4 **Bioinspired Silica for Enzyme Immobilisation: A Comparison with Traditional Methods** 39
Claire Forsyth and Siddharth V. Patwardhan

5 **On The Immobilization of *Candida antarctica* Lipase B onto Surface Modified Porous Silica Gel Particles** 63
Stephen J. Clarson, Richard A. Gross, Siddharth V. Patwardhan and Yadagiri Poojari

6 **Enzymatic Modification and Polymerization of Siloxane-Containing Materials** 73
Mark B. Frampton, Jacqueline P. Séguin and Paul M. Zelisko

7 **Design and Thermal Properties of Interpenetrating and Intercrosslinked Biosilicate Materials** 91
Andrew J. Vreugdenhil, Christophe Bliard, Shegufa Merchant and Suresh S. Narine

8 Bioactive Amino Acids, Peptides and Peptidomimetics Containing Silicon 103
Scott McN. Sieburth

Index 125

Contributors

C. Bliard Institut de Chimie Moléculaire de Reims, CNRS UMR7312 ICMR, Université de Reims, Reims, Cedex 2, France

S. J. Clarson Department of Chemical and Materials Engineering and the Polymer Research Centre, The University of Cincinnati, Cincinnati, OH, USA

C. Forsyth Department of Chemical and Process Engineering, University of Strathclyde, Glasgow, U.K.

M. B. Frampton Department of Chemistry and Centre for Biotechnology, Brock University, St. Catharines, ON, Canada

R. A. Gross Department of Chemistry and Biology, Rensselaer Polytechnic Institute, Troy, NY, USA

S. A. Guruswamy-Thangavelu Department of Chemistry, Trinity University, San Antonio, TX, USA

J. B. Lambert Department of Chemistry, Trinity University, San Antonio, TX, USA

S. Merchant Trent Centre for Biomaterials Research, Trent University, Peterborough, ON, Canada

S. S. Narine Department of Chemistry, Trent University, Peterborough, ON, Canada

Trent Centre for Biomaterials Research, Trent University, Peterborough, ON, Canada

Department of Physics and Astronomy, Trent University, Peterborough, ON, Canada

S. V. Patwardhan Department of Chemical and Process Engineering, University of Strathclyde, Glasgow, UK

Y. Poojari Department of Chemical and Materials Engineering and the Polymer Research Centre, The University of Cincinnati, Cincinnati, OH, USA

J. P. Séguin Department of Chemistry and Centre for Biotechnology, Brock University, St. Catharines, ON, Canada

S. McN. Sieburth Department of Chemistry, Temple University, Philadelphia, PA, USA

A. J. Vreugdenhil Department of Chemistry, Trent University, Peterborough, ON, Canada

Trent Centre for Biomaterials Research, Trent University, Peterborough, ON, Canada

P. M. Zelisko Department of Chemistry, Centre for Biotechnology, Brock University, St. Catharines, ON, Canada

Chapter 1
Silicon in a Biological Environment

Paul M. Zelisko

1.1 Silicon-Based Life: Science-Fiction?

Given that the periodic table groups atoms with similar properties and reactivities should silicon-based life not be possible? Since the late 1800s science fiction writers such as H.G. Wells [1] and Stanley Weinbaum [2] have imagined silicon-based forms of life.[1] More recently the original *Star Trek* series, in an episode entitled "Devil in the Dark" depicted an organism called a Horta that terrorized a colony of miners. However, in most instances the silicon-based organisms were simply rocks with a personality.

As much as it is intriguing to imagine organisms based on silicon, this does not seem to be possible, given our understanding for the requirements for life at the very least. Although both are members of Group IV, at a very fundamental level carbon and silicon are two different types of elements; carbon is a non-metal while silicon is classed as a metalloid. The true limitation for silicon in serving as the basis for life as we understand it is the inherent reactivity of the bonds that silicon makes with heteroatoms such as nitrogen and sulfur, bonds that tend to be relatively stable in carbon-based systems; these bonds are essential for many of the aspects of the biomachinary on Earth. A notable exception to this observation is the strength of the Si-O bond compared to the C-O bond (Table 1.1).

As can be seen from Tab. 1.1 C–C bonds are shorter and stronger than Si-Si bonds. Couple this with the fact that unlike carbon, silicon is incapable of making double and triple bonds with heteroatoms such as nitrogen and oxygen that are

[1] "The beast was made of silica! There must have been pure silicon in the sand, and it lived on that…It was silicon life!".[2]

P. M. Zelisko (✉)
Department of Chemistry, Centre for Biotechnology, Brock University,
500 Glenridge Avenue, St. Catharines, ON L2S 3A1, Canada
Tel.: +1-905-688-5550
e-mail: pzelisko@brocku.ca

P. M. Zelisko (ed.), *Bio-Inspired Silicon-Based Materials,* Advances in Silicon Science 5,
DOI 10.1007/978-94-017-9439-8_1

Table 1.1 Bond strengths and bond lengths for selected biologically relevant bonds to carbon and the corresponding bonds to silicon [3–5]

Bond	Bond energy (kJ/mol) [kcal/mol]	Bond length (Å)	Bond	Bond energy (kJ/mol) [kcal/mol]	Bond length (Å)
Si–C	369 [88]	1.89	C–C	334 [80]	1.54
Si–H	376 [90]	1.48	C–H	420 [100]	1.09
Si–O	531 [127]	1.63	C–O	340 [81]	1.41
Si–N	401 [96]	1.74	C–N	335 [80]	1.47
Si–S	414 [99]	2.14	C–S	313 [75]	1.80
Si–Si	308 [74]	2.34			

stable for a biologically significant timeframe and it becomes more and more difficult to imagine silicon-based life existing, at least according to our definition of life.

That being said, the first steps to silicon-based life may have been taken by researchers who have developed zeolite catalysts that behave as enzyme mimics. These systems, which support oxidizing Fe and Pd species, have the capacity to activate molecular oxygen and perform hydrocarbon oxidation in a manner similar to cytochrome P450 [6]. Also, recent research has shown that silicon may have played an integral role in the formation of what is now ribonucleic acid (RNA) molecules (Chap. 2). It may not be silicon-based life, but is it a step in that direction?

1.2 Not Science Fiction After All: Plants, Diatoms, and Sponges

As much as life as we know it is based on carbon rather than silicon, silicon does play a very important role in the life cycles of a number of marine and terrestrial organisms [7–10]. Of these the diatoms, sponges, and plants are perhaps the most well known [11].

1.2.1 Plants

Silicon has been reported to play a number of roles in plant species, such as the banana, [12] cucumber, [13, 14] sorghum, [15] wheat, [16] rice, [17] and horsetail, [18] which absorb silicon from the soil in the form of silicic acid [19]. Although the beneficial role of silicon in promoting the growth and development of a number of plant species is generally accepted, the exact physiological and metabolic mechanisms with these phenomena are not well understood and in some cases ignored altogether [20, 21].

In some plant species silicon is drawn into the plant and deposited in epidermal cells or as phytoliths in lumen cells [19]. In this role the silica that is deposited serves to strengthen the plant and to provide rigidity. The silica that is deposited in the phytoliths effectively acts as a skeletal system for the plant.

Not only does the silicon taken up by plants impart structural reinforcement for the plant, research has also demonstrated that silicon aids the plant in resisting biotic (e.g., insects, herbivores, bacteria, fungi) and abiotic (e.g., wind, cold, heat, salinity, droughts) stressors.[21] Related to this phenomenon is the roles that silicon plays in carbohydrate synthesis (Chap. 2) and phenolic synthesis, especially in rice [22].

Since plants lack the capacity for locomotion that allows animals to evade predators they have evolved a strategy of physical defense with which silicon plays an integral role. Physical features such as thorns, spines, and rough surface cell layers, all of which have been shown to be rich in silica, are all examples of strategies that plants have evolved to protect themselves against insect pests and herbivores [21, 23]. These physical characteristics afford plants physical barriers against the ingress of insects and pathogens while thorns, spines, and "chewy" exteriors may convince herbivores to seek easier meals elsewhere. Research has also demonstrated that the ingestions of silicon-rich plants by insects may limit the ability of the insects to ingest sufficient quantities of other nutrients and/or water [23, 24].

Silicon is also believed to be responsible for the production of secondary metabolites in plants as a response to biotic stressors [12, 21, 25]. These secondary metabolites are believed to help the plant defend itself against predators but also microorganisms. Studies have shown that some plants do contain genes whose expression is regulated by silicon [26]. Plants (*Arabidopsis*) infected with powdery mildew demonstrated a clear difference in the expression of certain genes and the expression of these genes was regulated by silicon. In control plants there was little difference in gene expression [27].

1.2.2 Marine Organisms

Perhaps the most widely studied groups of organisms that utilize silicon are found fresh and salt water systems. It has been estimated that the Earth's oceans contain 10^{17} moles of dissolved silicic acid ($Si(OH)_4$) at an average concentration of approximately 70 μM [28]. Given these relatively high values of $Si(OH)_4$ in the world's bodies of water it stands to reason that some aspect of Nature would evolve a use for such an abundant compound.

1.2.2.1 Diatoms

Diatoms are single-celled photosynthetic eukaryotes and constitute the largest class of protists (> 10,000 species) living in the planet's bodies of water [29–31]. A great deal of scientific effort has been committed to exploring how these organisms take in silicic acid from their environment and use it to construct ornate, species-specific silicon-based skeletal systems (Fig. 1.1) [32–35]. Diatoms utilize polycationic peptides

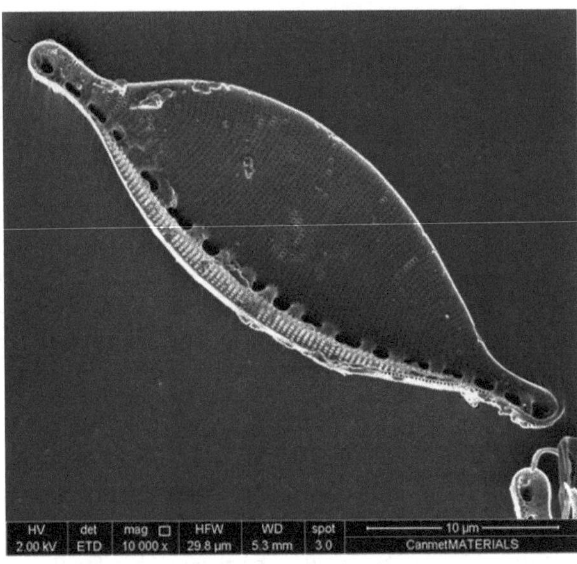

Fig. 1.1 Scanning electron micrograph of the diatom *Nitzshia curvilineata* [37, 38]

called *silaffins* to not only condense the silicic acid into the silica framework that comprises their skeletal systems, but to also template out their ornate structures [36].

One body of research has led the hypothesis that diatoms coevolved with their environments; the affinity of the active silicon uptake mechanism increased as silicon levels in marine environments decreased. It has even been hypothesized that silicon played a crucial role in providing a stable mineral surface on which the assembly and replication of primitive genetic material; [39] this concept is explored further in Chap. 2. Both of these theories may hold the key to understanding why diatoms evolved to process silicon compounds rather than carbonate like so many other marine organisms.

1.2.2.2 Sponges

A major breakthrough in the study of biogenic silica was made by Morse and coworkers in their studies of the marine sponge *Tethya aurantia* [40–43]. Approximately 75 % of the sponges dry mass is comprised of silica spicules that serve not only as a support system for the organism but also as a defense against predation [31]. Upon dissolution of the silica spicules using HF Shimizu et al. demonstrated that the spicules contained three (α, β, and γ) protein strands that were termed *silicatein* (*silica* pro*tein*) (Fig. 1.2) [40]. Analysis of silicatein α indicated that the protein possessed homology with the cathepsin L and papain family of proteases.

Perhaps one of the more illuminating observations to come from Morse's work was that if the silicatein isolated from the sponge spicules was challenged with tetraethoxysilane (TEOS, a surrogate for silicic acid), the protein could catalyze the hydrolysis of TEOS and condensation of the $Si(OH)_4$ to yield silica [41]. Native silicatein produced 214.0 ± 2.0 nmol of silica compared to denatured silicatein

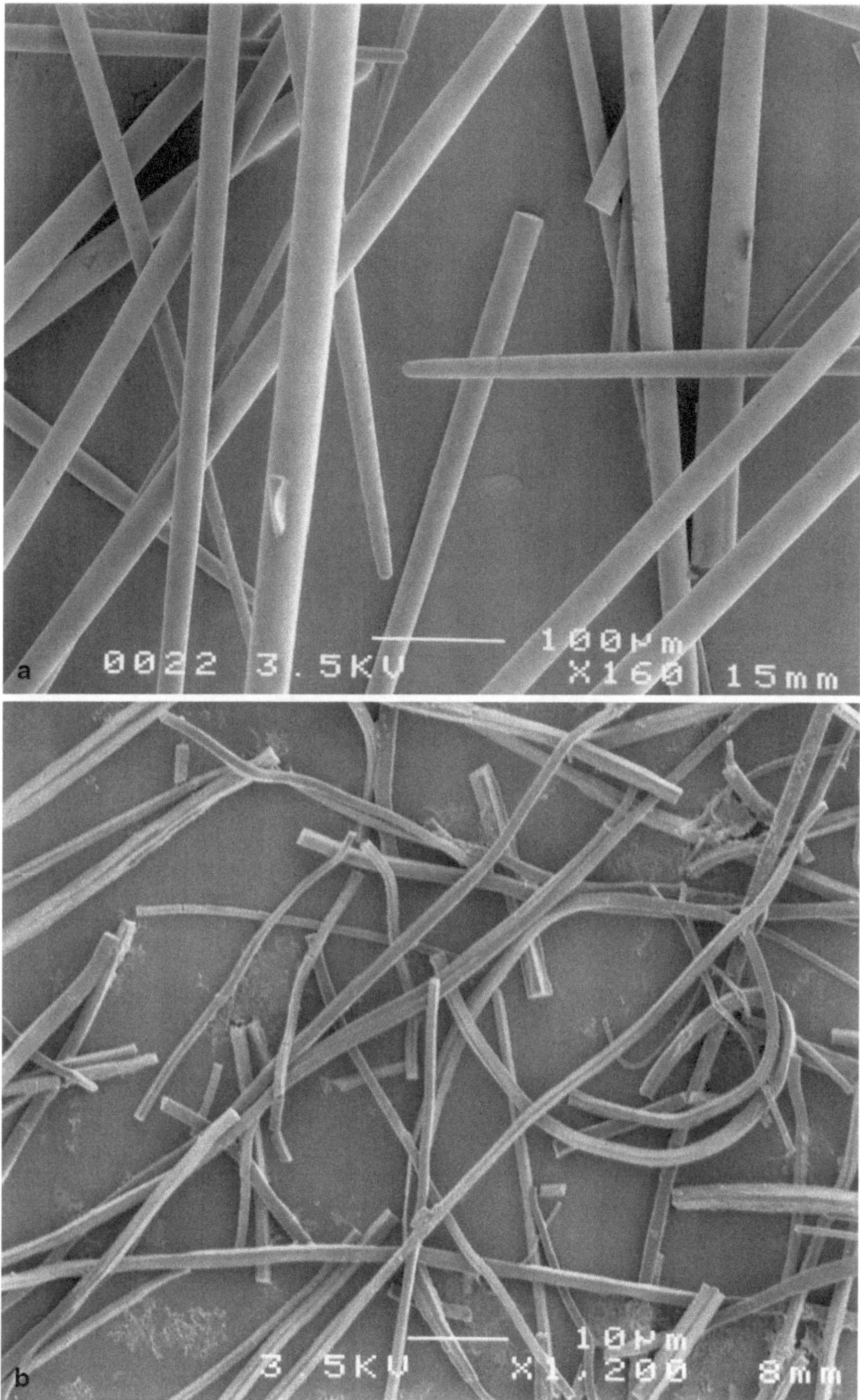

Fig. 1.2 Scanning electron micrographs of isolated silica spicules (× 130) (**a**) and axial filaments (× 1000) (**b**) from *Tethya aurantia*. [40] Reproduced with permission. Copyright 1998 National Academy of Sciences, U. S. A.

(24.5±2.0 nmol), bovine serum albumin (42.1±0.7 nmol), papain (22.9±1.0 nmol), trypsin (16.2±2.6 nmol), and protein-free (10.2±1.3 nmol) experiments. These data clearly indicated that the capacity to synthesize silica was not generalizable to all proteins but that a native silicatein structure was required in order observe this phenomenon in an efficient manner.

Attempts to gain an understanding of how diatoms processed silicic acid dissolved in water into ornate physical structures served as a starting point for forays into silicon biotechnology (Sect. 1.3, Chaps. 3, 5, and 6).

1.3 Drawing Inspiration from Nature

Given the natural interaction between certain silicon compounds with biological systems, it is perhaps not surprising that a great deal of research is concerned with incorporating and/or applying silicon-based compounds to biological systems in a controlled manner. There have been a great number of silicon-based compounds developed for applications in biologically relevant systems. To fully review the complete gamut of silicon compounds developed for biological systems is beyond the scope of this chapter. This chapter will provide only a very brief introduction into the application of silicon-based compounds to the biomedical and agricultural fields.

1.3.1 Applications in Agriculture

Aside from conferring resistance to pests (Sect 1.2.1), silicon-based compounds have been used as delivery vehicles in agricultural spray applications as a means of increasing the efficacy of the compound being applied [44]. These so called "superwetters", which include compounds that fall under the Silwet® brand, have a tendency to undergo spontaneous spreading when applied to surfaces such as foliage (Fig. 1.3) [45]. Organosilicone-based surfactants typically have low surface tensions at equilibrium, which have been reported as low as 20 mN m^{-1} [44].

The silicone superwetters facilitate leaf wetting by any solutions being applied to the foliage. This phenomenon permits herbicides to travel to the underside of leaves, which allows the spray solution to enter open stomata, which improves the uptake, and performance of herbicides [44]. It has been postulated that the silicone alkylenoxides commonly used as superwetters create a surface tension gradient which ultimately facilitates its spreading [45].

1.3.2 Silicon in Human Health and Medicine

Given the natural abundance of silicon on the planet (silicon makes up approximately 28% of the Earth's crust [46, 47]), it stands to reason that Nature has evolved to

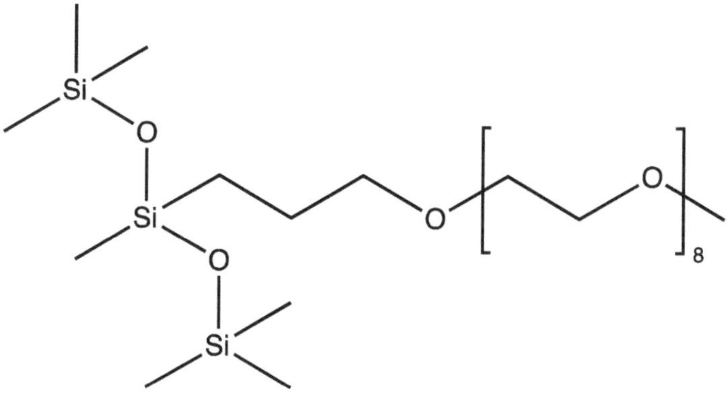

Fig. 1.3 The structure of Silwet L-77 [45]

make use of this readily available resource in its many organisms. Also, the biocompatibility of silicone polymer systems makes them ideal choices for various biomedical applications [3].

1.3.2.1 Silicon in Food and in the Human Body

A number of reports in the literature have examined the silicon content of a variety of foodstuffs ranging from instant food products to fresh fruits and vegetables [48–50].

Quite obviously from the data presented in Table 1.2, silicon is present in a wide variety of prepared and fresh food and drinks. However, in spite of its ubiquitous presence in the human diet, not all of the silicon is necessarily available to us. For instance, a study by Robberecht et al. of Belgian foodstuffs reported that although whole rice has a silicon content of 162.0 mg/kg, only 7.6 ± 1.3 % of that silicon is available. Conversely, chicken breasts were found to have a silicon content of only 1.09 mg/kg, and of that amount 40 ± 4 % of the silicon was bioavailable [49]. It would appear that there is a great deal of variation not only between the silicon content of various foodstuffs, although perhaps not surprisingly plant-derived foods tend to have higher silicon contents, but also in the bioavailability of the silicon that foods do contain.

It stands to reason that if silicon is so prevalent in our food that it is likely to play a role in human physiology (and that of other vertebrates [51, 52]). In fact silicon has been shown to be a necessary component of bone growth and cartilage formation and may be a component of the system needed for the crosslinking of connective tissue [53–57]. In fact, one study suggests that beer may be a major contributor to silicon levels in the bodies of those who consume the beverage [58]. Analysis demonstrated that *in vivo* silicon concentrations increased by as much as 713 ± 121 μg/L in subjects 1.5 h following the consumption of beer.

Table 1.2 The silicon content of some foods in the United Kingdom. Adapted from Reference 49 with permission. © Cambridge University Press, 2005

Food description	Portion (g)	mg/portion	Mean mg/100g
Chocolate-covered digestive cookies	2 cookies (36)	0.88	2.44
All-butter shortbread cookies	2 cookies (26)	0.31	1.19
Bread, white	1 slice (36)	0.64	1.79
Bread, whole wheat	1 slice (36)	1.72	4.78
Couscous	3 tablespoons (99)	0.54	0.55
Flour, white	1 tablespoon (20)	0.86	4.29
Flour, whole wheat	1 tablespoon (20)	0.61	3.04
Oat bran	2 tablespoons (14)	3.27	23.36±8.88
Porridge oats (dry)	2 tablespoons (30)	3.42	11.39
Rice, arborio	3 heaped tablespoons (120)	1.06	0.88
Wheat bran	2 tablespoons (14)	1.54	10.98±9.03
Apples (eating), raw	1 medium without core (100)	0.21	0.21
Apricots, fresh	1 medium without stone (40)	0.44	1.11
Banana, raw	1 medium without skin (100)	4.77	4.77
Cherries, raw	10 cherries without stone (40)	0.41	1.03
Clementine, raw	1 medium without skin (60)	1.51	2.52
Grapes (green and red)	1 small bunch (100)	0.49	0.49
Strawberries, raw	10 pieces (120)	1.19	0.99
Baked beans, canned in tomato sauce	1 tablespoon (40)	0.37	0.92
Hummus	1 tablespoon (30)	0.29	0.97
Tofu	1 average serving (60)	1.78	2.96
Cheese, hard	1 medium piece (40)	0.19	0.47
Cheese, soft	1 medium piece (40)	0.16	0.39

Table 1.2 (continued)

Food description	Portion (g)	mg/portion	Mean mg/100g
Milk, partially skimmed	1 glass (200)	0.14	0.07
Beetroot, pickled, drained	4 slices (40)	0.32	0.79
Carrots, raw	1 tablespoon (40)	0.04	0.10
Celery, raw	1 stalk (30)	0.09	0.29
Cucumber, raw	2.54 cm piece (60)	1.52	2.53
Iceberg lettuce, raw	1 average serving in a salad (80)	0.54	1.81
Potato, new, peeled, boiled	1 medium (40)	0.22	0.56
Radish	5 medium (40)	0.16	0.39
Spinach, fresh, boiled	2 tablespoons (80)	4.10	5.12
Sweet corn, on the cob, boiled	1 cob, kernels only (125)	0.48	0.38
Coffee, instant	1 mug (260)	1.53	0.59
Cola, canned	1 can (343)	0.69	0.20
Tea, black, tea bag	1 mug (260)	0.31	0.86
Bottled water	1 bottle (330)	1.65	0.50
Tap water	1 glass (200)	0.50	0.25
Bitter or ale, draft	1 pint (547)	12.60	2.19±0.60
Cider	1 pint (574)	2.30	0.40±0.07
Lager, draft	1 pint (574)	16.30	2.84±1.05
Stout, bottled	1 bottle (300)	5.28	1.76±0.12
Port	1 small glass (50)	0.62	1.24±0.15
Red wine	1 glass (125)	0.85	0.68±0.59
White wine	1 glass (125)	1.34	1.07±0.47

In studying the serum silicon levels in apparently healthy individuals, Bissé et al. noted differences between men and women that were significantly age dependent [55]. In study populations that were 18–44 years of age women were found to have a higher serum silicon level than men up to a maximum difference of 21 % which was observed in individuals in the 30–44 years old age group. This study noted that as men age from 18–59 years of age their serum silicon levels increase to a median of 9.7–10.2 μmol/L. As women aged from 18–44 years of age their serum silicon levels were shown to increase with a median range of 10.0–11.1 μmol/L. Over the age of 60 women exhibited a decrease in their serum silicon concentrations of approximately 28 %. Depending on the analytical technique employed, other studies have reported human serum and plasma silicon values ranging from 0.60 ± 0.36 to 21.5 ± 4.5 μmol/L [59–70] Brown reported that one area where the human body tends to exhibit a decrease in silicon content is the skin [71]. Based on the observed serum silicon levels in women it has been speculated that serum silicon levels correlate with hormone levels as women in age groups that had the highest hormonal activity also demonstrated the highest level of silicon concentrations [55, 72].

1.3.2.2 Silicon in Medicine

1.3.2.2.1 Silicon-Containing Molecules with Medicinal Applications

A number of silicon-containing molecules have been synthesized in the last 30 years or so with designs on somehow influencing animal physiology. The use of silicon containing small molecules as bioactive compounds has garnered a great deal of attention because in many cases the differences in the chemical properties of the silicon compounds demonstrate enhanced potency and improved pharmocological characteristics in comparison to the carbon analogues (Chap. 8) [73].

A number of silicon-containing drug compounds have been inspired by purely organic drugs or as variants of existing organic-based drug compounds (Fig. 1.4) [73–89].

Cyclooxygenase-2 (COX-2) has been implicated in a number of pathological conditions. Non-steroidal anti-inflammatory drugs (NSAIDs) have demonstrated effectiveness at inhibiting COX, however, in addition to inhibiting COX-2 NSAIDs also inhibit COX-1 which is involved in renal and gastrointestinal protection and may cause a number of drug-related side effects [90]. As a result finding a selective COX-2 inhibitor has garnered a great deal of attention. Compound **1** is a lipophilic silicon-based analogue of indomethacin and has proven to selective for inhibiting human COX-2 in recombinant cell lines [74, 80]. The derivative of **1** where $R^1 = R^2 = R^3 = Me$ and $n = 3$ demonstrated antiproliferative activity against pancreatic carcinoma cells with an $IC_{50} = 6.0$ μM compared to an $IC_{50} > 100$ μM for indomethacin itself.

Phthalocyanine 4 (Pc 4) **2** developed by Kenney et al. represents a potential tool in the photodynamic therapy of cancer by employing a photosensitizer (in this case **2**), light and oxygen to eradicate cancer cells [77, 82–84]. Pc 4 **2** has demonstrated encouraging anti-cancer activity both *in vitro* and *in vivo* to the point where the molecule has entered clinical trials [83]. However, given the hydrophobicity

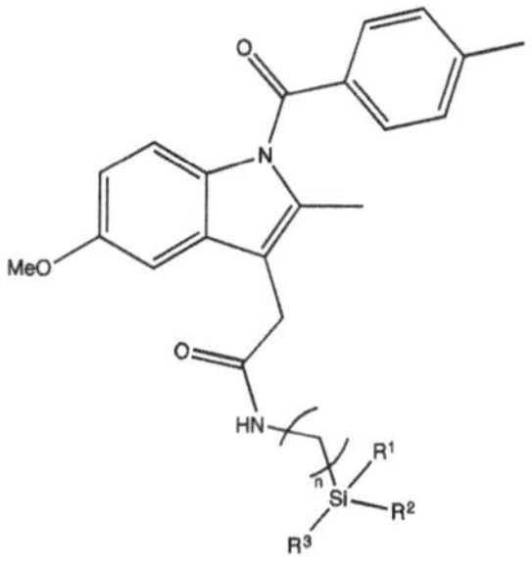

Fig. 1.4 Examples of silicon-containing drug compounds

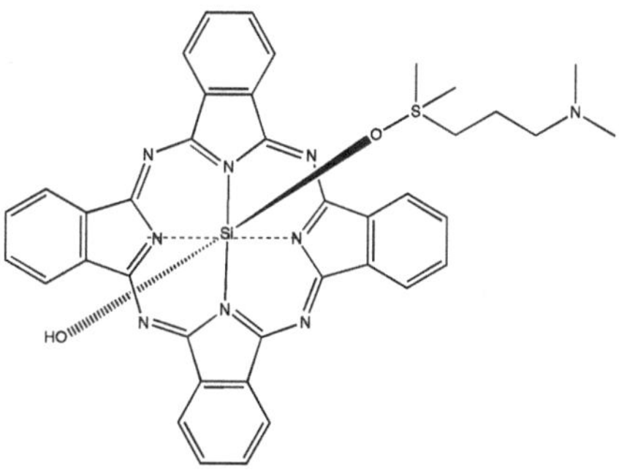

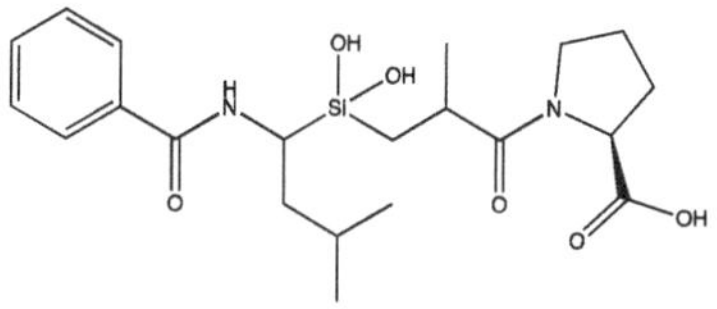

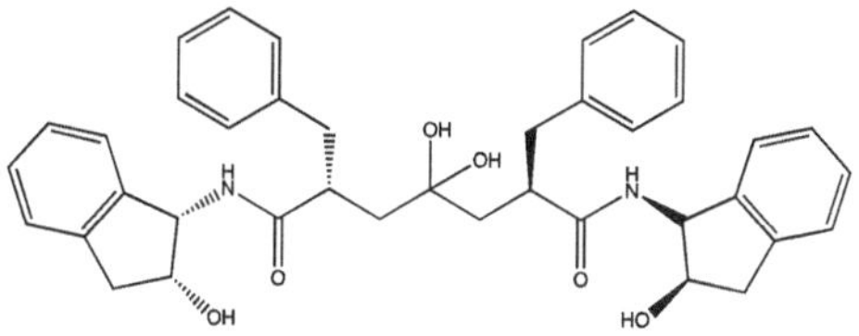

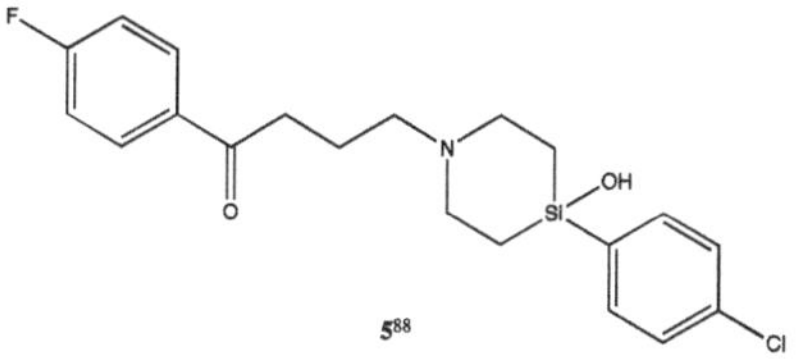

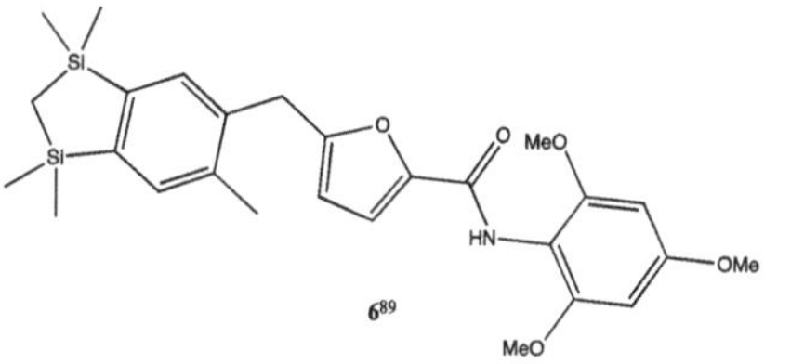

of Pc 4 **2** block copolymer micelles have been explored as means of delivering the drug *in vivo* [77].

Compounds **3** and **4** represent a class of molecules designed to behave as bioisosteres of stabilized protease acyl-enzyme intermediates thereby inhibiting the activity of those enzymes; the molecule enters the enzyme active site and effectively binds in that position limiting access to the active site by the natural substrate for the enzyme [75, 78, 79]. Compound **3**, which targets angiotensin converting enzyme (ACE), has demonstrated an $IC_{50}=14$ nM while **4** has an $IC_{50}=2.7$ nM when challenged with HIV protease [78].

Haloperidol is a dopamine receptor antagonist used in the treatment of neurological disorders such as schizophrenia and Parkinson's disease. However, in addition to its clinical uses haloperidol also exhibits neurotoxic side effects [88]. It has been postulated that a pyridinium metabolilte of haloperidol is responsible for the observed neurotoxic behavior of the drug because of its close structural resemblance to 1-methyl-4-phenylpyridinium which has been shown to induce Parkinson-like symptoms [91, 92]. The sila-haloperidol **5** analogue of haloperidol does not form a

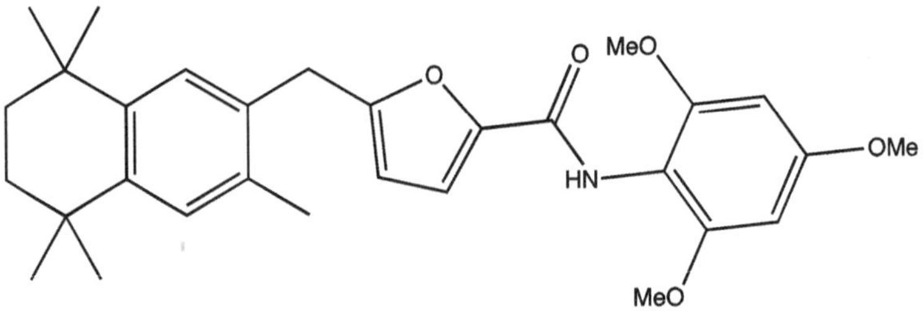

Fig. 1.5 GnRH antagonist AG-045572 [89]

sila-pyridinium ion as a metabolite. As a result the sila-haloperiodol may provide an alternative treatment modality for a number of neurological disorders without the negative side effects observed with haloperidol.

Drug molecules that act as antagonists and agonists for gonadotropin-releasing hormones (GnRH) are being explored as a means of treating various reproductive disorders and tumors that may be linked to hormones [93]. One class of compounds that is being examined for these applications are molecules that are not based on peptides but still have the potential to behave as GnRH agonists or antagonists; 5-[(1,1,3,3,6-pentamethyl-1,3-disila-2,3-dihydro-1*H*-inden-5-yl)methyl]-*N*-(2,4,6-trimethoxyphenyl)furan-2-carboxamide **6** falls into this category [89]. A sila-analogue of the non-peptidic GnRH antagonist AG-045572 (Fig. 1.5) **2** demonstrated a retention of efficacy in spite of the insertion of silicon atoms into one of the ring systems of the AG-045572 structure [89]. The insertion of silicon atoms into key positions in existing drug compounds may present pharmaceutical researchers with access to new libraries of bioactive compounds as well as avenues around the limitations placed on drug discovery by the existing patent literature.

Few silicon-containing drug candidates have progressed to clinical trials [3, 80, 94]. One example of a silicon-based drug that is an exception to this statement is 7-[(2-trimethylsilyl)ethyl]-20(*S*)-camptothecin (BNP1350 or karenitecin) (Fig. 1.6) [85, 86]. An analogue of camptothecin, which is isolated from the *Camptotheca acuminata* tree, karenitecin has demonstrated a broad spectrum of activity against experimental human tumors and is a candidate for the oral treatment of cancer. Karenitecin's potency, broad spectrum of activity, oral bioavailability, lactone stability, lack of metabolic conversion, and chemical stability make it an attractive candidate as a chemotherapy agent [86]. As of 2004 karenitecin was progressing through Phase I and Phase II clinical trials [85].

Germanium-containing bioactive compounds have also been reported based on C/Si/Ge bioisoterism [96].

1.3.2.2.2 Biomedical Devices Based on Silicon

One of the areas that silicon species have received a great deal of attention has been in the area of biomedical applications. The following sections will provide a very

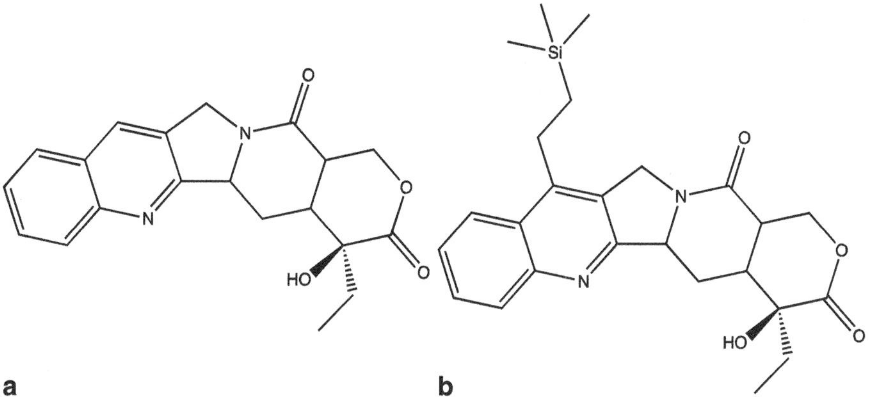

Fig. 1.6 Structural formulae of camptothecin [95] (**a**), and karenitecin (**b**) [85, 86]

brief glimpse at some of the roles that silicon-based species have played and continue to play in the biomedical field. A comprehensive treatment of this field could undoubtedly be the subject of several texts in its own right.

Related to the use of silicon-containing bioactive molecules (Sect. 1.3.2.2.1) silicone matrices have been explored for the controlled release of biologically active molecules since the 1960s typically as matrix or reservoir devices [97]. For example, Norplant® is a sub-cutaneous implant designed to release the contraceptive levonorgestrel over a 5-year period [98]. A related system, Estring®, is a vaginal insert that is used to treat the symptoms of menopause by releasing 17-β-estradiol [99]. More recently silicone hydrogel contact lenses were doped with levofloxacin (an antibiotic) and chlorhexidine (an antiseptic) for delivery of these materials directly to the eye. Given the successful release of these materials from the contact lens the development of a daily disposable therapeutic contact lens may be possible [100].

In addition to the polymeric silicone systems, porous silicon has also garnered attention as drug delivery vectors because of their biocompatibility, biodegradability, ease of fabrication, their tunable structure, and the porous nature of the network [101, 102]. These types of materials have been used to deliver localized radioactive ^{32}P to tumors. Chemotherapy agents such as cisplatin and doxorubicin have also been delivered from porous silicon *in vitro* [103, 104].

In addition to its uses as a drug delivery vehicle, porous silicon has also received attention as a platform onto which various tissues can adhere and ultimately grow [105, 106]. Voelcker and colleagues have utilized porous silicon as a non-toxic, biodegradable scaffold onto which they have promoted the adhesion of cells related to the human eye. Similarly, silicon is often incorporated into medical implants and bone grafts to promote adhesion between the implant and the existing bone tissue [107]. Similar research has looked at using porous silicon as a means of effecting soft and cartilage tissue repair [108].

Silicone polymers based on a Si-O-Si backbone are perhaps the widely known and widely applied silicon-based system in biomedical devices. A number of reasons account for this observation [3, 109]:

Table 1.3 Examples of *in vivo* applications of silicone polymers [3, 110–112]

Body region	Biomedical device
Head	Ear/nose cartilage replacement
	Replacement for vitreous humour in the eye
	Hydrocephalus shunt
	Dura membrane
	Eustachian tube
	Tracheal stent
	Tracheostomy vent
	Synthetic eyeball
	Orbital floor implant
	Mandibular prosthesis
Torso	Breast implants
	Penile implants
	Heart valves
	Pacemaker leads
	Artificial skin
	Anti-reflux cuff
	Urethral cuffs
	Vaginal stents
	Ureteral stent
	Oviductal plugs
	Testicular prosthesis
Upper limb	Finger joints
	Wrist joints
	Artificial nail
	Elbow joint cap
Lower limb	Knee joints
	Hip joint impact cup
	Toe joints

- silicones have a very low glass transition temperature (146 K) when compared to their hydrocarbon analogues (200 K for polyisobutylene)
- gases (oxygen, nitrogen, and water vapor) have a high degree of solubility and high diffusion coefficient in silicones
- silicones have a great deal of compressibility
- silicones are relatively bioinert
- silicones possess properties that cannot be achieved with purely organic polymers

Silicone polymers, typically as elastomers, have found utility in a number of *in vivo* applications as illustrated in Table 1.3. Silicone biomaterials were first developed in the 1950s with a shunt designed to relieve pressure on the brains of children afflicted with hydrocephalus [110].

However, the availability of silicone biomaterials has declined in recent years as many suppliers have either ceased producing silicone biomaterials or severely limited their distribution as a result of concerns resulting from product liability and potential litigation, particularly in the United States [113]. For a number of implantable materials still on the market limitations are placed on how long the material can remain within the body; Dow Corning Corporation limits the implantation of its silicone materials to 29 days [3].

References

1. Wells HG (1894) Saturday Rev 676
2. Weinbaum SG (2008) A martian odyssey: Stanley G. Weinbaum's worlds of if. Wildeside Press, p 96
3. Brook MA (2000) Silicon in organic, organometallic, and polymer chemistry. Wiley, New York
4. Doncaster AM, Walsh R (1979) J Chem Soc Chem Commun 904
5. Colvin EW (1988) Silicon reagents in organic synthesis. Academic Press, London p 1
6. Herron N (1989) Zeolite catalysts as enzyme mimics: toward silicon-based life? Biocatal Biomim (ACS Symp Ser) 392:141–154
7. Leng MJ, Swann GEA, Hodson MJ, Tyler JJ, Patwardhan SV, Sloane HJ (2009) Silicon 1:65–77
8. Schroeder HC, Brandt D, Schlossmacher U, Wang X, Tahir MN, Tremel W, Belikov SI, Mueller WEG (2007) Naturwissenschaften 94:339–359
9. Coradin T, Lopez PJ (2003) Chem Bio Chem 4:251–259
10. Sangster AG, Hodson MJ, Tubb HJ (2001) Stud Plant Sci 8:85–113
11. Schröder HC, Wang X, Tremel W, Ushijima H, Müller WEG (2008) Nat Prod Rep 25:455–474
12. Vermeire M-L, Kablan L, Dorel M, Delvaux B, Risède J-M, Legrève A (2011) Eur J Plant Pathol 131:621–630
13. Faisal S, Callis KL, Slot M, Kitajima K (2012) Botany 90:1058–1064
14. Correa RS, Moraes JC, Auad AM, Carvalho GA (2005) Neotrop Entomol 34:429–433
15. Costa RR, Moraes JC (2002) Ecossistema 27:37–39
16. Basagli MA, Moraes JC, Carvalho GA, Ecole CC, Goncalves-Gervasio R de CR (2003) Neotrop Entomol 32:659–663
17. Chen W, Yao X, Cai K, Chen J (2011) Biol Trace Elem Res 142:67–76
18. Law C, Exley C (2011) BMC Plant Biol 11:112–120
19. Vasanthi N, Saleena LM, Anthoni RS (2012) World Appl Sci J 17:1425–1440
20. Detmann KC, Araujo WL, Martins SCV, Fernie AR, Da Matta FM (2013) Plant Signal Behav 8:e22523/71–e22523/74
21. Epstein E (2009) Ann Appl Biol 155:155–160
22. Van Soest PJ (2006) Anim Feed Sci Technol 130:137–171
23. Reynolds OL, Keeping MG, Meyer JH (2009) Ann Appl Biol 155:171–186
24. Massey FP, Ennos AR, Hartley SE (2006) J Anim Ecol 75:595–603
25. Van Bockhaven J, De Vleesschauwer D, Höfte M (2013) J Exp Bot 64:1281–1293
26. Watanabe S, Shimoi E, Ohkama N, Hayashi H, Yoneyama T, Yazaki J, Fujii F, Shinbo K, Yamamoto K, Sakata K, Sasaki T, Kishimoto N, Kiuchi S, Fujiwara T (2003) Soil Sci Plant Nutr 50:1273–1276
27. Fauteux F, Chain F, Belzile F, Menzies JG, Bélanger RR (2006) Proc Natl Acad Sci U S A 103:17554–17559

28. Tréguer P, Nelson DM, Van Bennekom AJ, DeMaster DJ, Leynaert A, Quéguiner B (1995) Science 268:375–379
29. Lopez PJ, Desclés J, Allen AE, Bowler C (2005) Curr Opinion Biotechnol 16:180–186
30. Ezzati J, Dolatabadi N, de la Guardia M (2011) Trends Anal Chem 30:1538–1548
31. Tacke R (1999) Angew Chem Int Ed 38:3015–3018
32. Hildebrand M (2008) Chem Rev 108:4855–4874
33. Hildebrand M (2003) Prog Org Coatings 47:256–266
34. Sumper M, Brunner E (2008) Chem Bio Chem 9:1187–1194
35. Brunner E, Gröger C, Lutz K, Richthammer P, Spinde K, Sumper M (2009) Appl Microbiol Biotechnol 84:607–616
36. Kröger N, Deutzmann R, Sumper M (1999) Science 286:1129–1132
37. DeJong JL (2013) Exploring the silicon substrate tolerance of the diatom nitzshia curvilineata. Undergraduate Thesis, Brock University, Department of Chemistry, St. Catharines, Ontario, Canada
38. The author would like to thank Renata Zavadil of Canmet Materials, Hamilton, Ontario, Canada for her assistance in acquiring the SEM image
39. Trevors JT (1997) Antonie van Leeuwenhoek 71:271–276
40. Shimizu K, Cha J, Stucky GD, Morse DE (1998) Proc Natl Acad Sci U S A 95:6234–6238
41. Cha JN, Shimizu K, Zhou Y, Christiansen SC, Chmelka BF, Stucky GD, Morse DE (1999) Proc Natl Acad Sci U S A 96:361–365
42. Zhou Y, Shimizu K, Cha JN, Stucky GD, Morse DE (1999) Angew Chem Int Ed 38:780–782
43. Morse DE (1999) Trends Biotechnol 17:230–232
44. Knoche M (1994) Weed Res 34:221–239
45. Nikolov AD, Wasan DT, Chengara A, Koczo K, Policello GA, Kolossvary I (2002) Adv Colloid Int Sci 96:325–338
46. Lutgens FK, Tarbuck EJ (2000) Essentials of geology, 7th edn. Prentice Hall, Upper Saddle River, NJ
47. Greenwood NN, Earnshaw A (1984) Chemistry of the Elements. Pergamon Press, Oxford (Chapter 9, p 381)
48. Prescha A, Zablocka-Slowinska K, Jojka A, Grajeta H (2012) Food Chem 135:1756–1761
49. Robberecht H, Van Cauwenbergh R, Van Vlaslaer V, Hermans N
50. Powell JJ, McNaughton SA, Jugdaohsingh R, Anderson SHC, Dear J, Khot F, Mowatt L, Gleason KL, Sykes M, Thompson RPH, Bolton-Smith C, Hodson MJ (2005) Brit J Nutr 94:804–812
51. Carlisle EM (1972) Science 78:619–621
52. Schwartz K, Miline T (1992) Nature 239:333–334
53. Carlisle EM (1982) Nutr Rev 40:193–198
54. Carlisle EM (1970) Science 167:279–280
55. Bissé E, Epting T, Beil A, Lindinger G, Lang H, Wieland H (2005) Anal Biochem 337:130–135
56. Sripanyakorn S, Jugdaohsingh R, Thompson RPH, Powell JJ (2005) Nutr Bull 30:222–230
57. Maehira F, Iinuma Y, Eguchi Y, Miyagi I, Teruya S (2008) J Bone Miner Metab 26:446–455
58. Sripanyakorn S, Jugdaohsingh R, Elliot H, Walker C, Mehta P, Shoukru S, Thompson RPH, Powell JJ (2004) Brit J Nutr 91:403–409
59. Lo DB, Christian GD (1978) Microchem J 23:481–487
60. Mauras Y, Riberi P, Cartier F, Allain P (1980) Biomedicine 33:228–230
61. Dobbie JW, Smith MB (1982) Scott Med J 27:17–19
62. Berlyne GM, Caruso C (1983) Clin Chim Acta 129:239–244
63. Berlyne GM, Dudek E, Adler AJ, Rubin JE, Seidman M (1985) Kidney Int 28:175–177
64. Tanaka T, Hayashi Y (1986) Clin Chim Acta 156:109–113
65. Berlyne GM, Adler AJ, Ferran N, Bennett S, Holt J (1986) Nephron 43:5–9
66. Gittelman HJ (1990) J Anal Atom Spectrom 5:687–689
67. Roberts NB, Williams P (1990) Clin Chem 36:1460–1465
68. Gittelman HJ, Alderman F, Perry SJ (1992) Kidney Int 42:957–959
69. Teuber SS, Saunders RL, Halpern GM (1995) Biol Trace Elem Res 48:121–130
70. Leung FY, Edmond P (1977) Clin Biochem 30:399–403

71. Brown H (1927) J Biol Chem 75:789–794
72. Jugdaohsingh R, Anderson SHC, Tucker KL, Elliott H, Kiel DP, Thompson PH, Powell JJ (2002) Am J Clin Nutr 75:887–893
73. Franz AK, Wilson SO (2013) J Med Chem 56:388–405
74. Bikzhanova GA, Toulokhonova IS, Gately S, West R (2005) Silicon Chem 3:209–217
75. Sieburth SM, Nittoli T, Mutahi AM, Guo L (1998) Angew Chem Int Ed 37:812–814
76. Troegel D, Möller F, Tacke R (2010) J Organomet Chem 695:310–313
77. Master AM, Rodriguez ME, Kenney ME, Oleinick NL, Gupta AS (2010) J Pharm Sci 99:2386–2398
78. Showell GA, Mills JS (2003) Drug Discov Today 8:551–556
79. Mills JS Showell GA (2004) Expert Opin Investig Drugs 13:1149–1157
80. Gately S, West R (2007) Drug Develop Res 68:156–163
81. Min GK, Hernández D, Skrydstrup T (2013) Account Chem Res 46:457–470
82. Colussi VC, Feyes DK, Mulvihill JW, Li Y-S, Kenney ME, Elmets CA, Oleinick NL, Mukhtar H (1999) Photochem Photobiol 69:236–241
83. Miller JD, Baron ED, Scull H, Hsia A, Berlin JC, McCormick T, Colussi V, Kenney ME, Cooper KD, Oleinick NL (2007) Toxicol Appl Pharmacol 224:290–299
84. Rodriguez ME, Zhang P, Azizuddin K, Delos Santos GB, Chiu S-M, Xue L-Y, Berlin JC, Peng X, Wu H, Lam M, Nieminen A-L, Kenney ME, Oleinick NL (2009) Photochem Photobiol 85:1189–1200
85. Thompson PA, Berg SL, Aleksic A, Kerr JZ, McGuffey L, Dauser R, Nuchtern JG, Hausheer F, Blaney SM (2004) Cancer Chemother Pharmacol 53:527–532
86. Van Hattum AH, Pinedo HM, Schlüper HMM, Hausheer FH, Boven E (2000) Int J Cancer 88:260–266
87. Warneck JB, Cheng FHM, Barnes MJ, Mills JS, Montana JG, Naylor RJ, Ngan M-P, Wai M-K, Daiss JO, Tacke R, Rudd JA (2008) Toxicol Appl Pharm 232:369–375
88. Johansson T, Widolf L, Popp F, Tacke R, Jurva U (2010) Drug Metab Dispos 38:73–83
89. Barnes MJ, Burschka C, Büttner M, Conroy R, Daiss JO, Gray IC, Hendrick AG, Tam LH, Kuehn D, Miller DJ, Mills JS, Mitchell P, Montana JG, Muniandy PA, Rapely H, Showell GA, Tebbe D, Tacke R, Warneck JBH, Zhu B (2011) Chem Med Chem 6:2070–2080
90. Warner TD, Giuliano F, Vojnovic I, Bukasa A, Mitchell JA, Vane JR (1999) Proc Natl Acad Sci U S A 96:7563–7568
91. Subramanyam B, Woolf T, Catagnoli N Jr (1991) Chem Res Toxicol 4:123–128
92. Dauer W, Przedborski S (2003) Neuron 39:889–909
93. Betz SF, Zhu Y-F, Chen C, Struthers RS (2008) J Med Chem 51:3331–3348
94. Bains W, Tacke R (2003) Curr Opin Drug Discov Dev 6:526–543
95. Tanizawa A, Kohn KW, Kohlhagen G, Leteurtre F, Pommier Y (1995) Biochemistry 34:7200–7206
96. Tacke R, Heinrich R, Kornek T, Merget M, Wagner SA, Gross J, Keim C, Lambrecht G, Mutschler E, Beckers T, Bernd M, Reissmann T (1999) Phosphorus, Sulfur, Silicon 150-151:69–87
97. Aguadisch L, Colas A (1997) Chim Nouv 15:1779–1788
98. Sam AP (1992) J Control Release 22:35–46
99. Schmidt G, Andersson S-B, Nordle Ö, Johansson C-J, Gunnarsson PO (1994) Gynecol Obstet Invest 38:253–260
100. Paradiso P, Galante R, Santos L, Alves Matos AP, Calaço R, Serro AP, Saramago B (2014) J Biomed Mater Res Part B: Appl Biomater 102B:1170–1180
101. Chiappini C, Tasciotti E, Serda RE, Brousseau L, Liu X, Ferrari M (2011) Phys Status Solidi C 8:1826–1832
102. Anglin EJ, Cheng L, Freeman WR, Sailor MJ (2008) Adv Drug Deliv Rev 60:1266–1277
103. Vaccari L, Canton D, Zaffaroni N, Villa R, Tomen M, di Fabrizio E (2006) Microelectron Eng 83:1598–1601
104. Li X, Coffer JL, Chen Y, Pinizotto RF, Newey J, Canham LT (1998) J Am Chem Soc 120:11706–11709

105. Low SP, Williams KA, Canham LT, Voelcker NH (2006) Biomaterials 27:4538–4546
106. Low SP, Voelcker NH, Canham LT, Williams KA (2009) Biomaterials 30:2873–2880
107. Porter AE (2006) Micron 37:681–688
108. Gencer ZA, Odabas S, Sasmazel HT, Piskin E (2012) J Bioactive Compat Polym 27:419–428
109. Colas A, Curtis J (2004) Silicone biomaterials: history and chemistry. In: Ratner BD, Hoffman AS, Schoen FJ, Lemons JE (eds) Biomaterials science: an introduction to materials in medicine. p 80–86
110. Yoda R (1998) J Biomater Res Polym Edn 9:561–626
111. Kannan RY, Salacinski HJ, Ghanavi J, Narula A, Odlyha M, Peirovi H, Butler PE, Seifalian AM (2007) Plast Reconstr Surg 119:1653–1662
112. Lane TH, Burns SA (1996) Curr Topics Microbiol Immunol 210:3–12
113. Lee GM (1995) Med Device Technol 6:20–25

Chapter 2
The Role of Silicates in the Synthesis of Sugars Under Prebiotic Conditions

Joseph B. Lambert, Senthil Andavan Guruswamy-Thangavelu

2.1 Sugars and Life

Before life, there must have been the molecules of life—amino acids for proteins, heterocyclic bases such as adenine for nucleic acids, and sugars for polysaccharides. In 1953, Miller demonstrated that amino acids could be produced from mixtures of water, carbon monoxide, ammonia, and hydrogen with the help of an electric spark [1]. Oró and Kimball were able to prepare the nucleotide base adenine from hydrogen cyanide and ammonia [2]. Many advances on these experiments have been made in the subsequent decades. In 1967 Gabel and Ponnamperuma reported experiments indicating that simple sugars could be prepared from formaldehyde, based on the Butlerov reaction [3]. Breslow in 1959 already had provided an aldol mechanism for this reaction [4]. Reid and Orgel, however, in a companion paper to that of Gabel and Ponnamperuma, concluded that the Butlerov reaction was impractical because the sugars decompose and "some method of stabilizing the sugars is essential" [5]. The authors respectively used aluminosilicates and a mixture of carbonate and hydroxyapatite, but obtained yields under 5 % of unstable sugars.

2.2 The Formose Reaction

The Butlerov reaction (Fig. 2.1) had come to be known as the formose reaction, as it converts formaldehyde to sugars (-oses). Our expansion of the Breslow mechanism is as follows.

Formaldehyde (C1) alone is incapable of undergoing an aldol reaction, since it lacks an acidic proton alpha to the carbonyl group. The two-carbon sugar glycoladehyde (C2) is the smallest such system with an active hydrogen, so Breslow hypothesized it must be present in catalytic amounts. The product of the aldol condensation

J. B. Lambert (✉) · S. A. Guruswamy-Thangavelu
Department of Chemistry, Trinity University, One Trinity Place, San Antonio, TX 78212, USA
e-mail: jlambert@northwestern.edu

P. M. Zelisko (ed.), *Bio-Inspired Silicon-Based Materials,* Advances in Silicon Science 5,
DOI 10.1007/978-94-017-9439-8_2

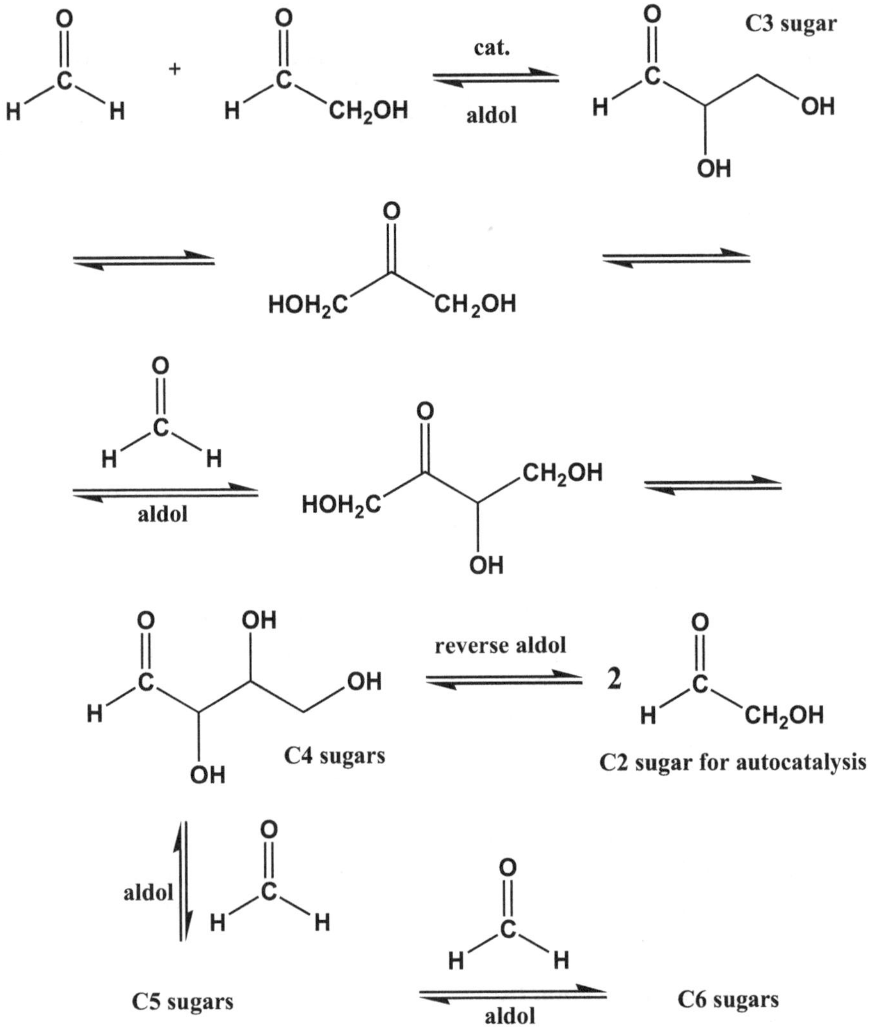

Fig. 2.1 The formose reaction

of C1 and C2 is glyceraldehyde (C3). Isomerization of glyceraldehyde to its ketose isomer and condensation of the ketose with a second molecule of C1 gives a four-carbon ketose, which can isomerize to the C4 aldose. Reverse aldol condensation of the C4 aldose delivers two molecules of the original catalyst, C2. Thus the reaction produces its own catalyst, i.e., it is autocatalytic. Following an induction period as the catalyst builds up, such a reaction continues exponentially. The C4 product also could add another molecule of formaldehyde to produce C5 products, which could condense again to form C6 products. Hence the products are multifarious and unstable, but are formed quickly and efficiently.

2.3 The Interaction(s) of Carbohydrates with Silicates

Our focus in this article is on carbohydrates, which constitute the most abundant organic materials in the biosphere, and in particular their interactions with silicates, the most abundant materials in the lithosphere. Although no naturally occurring organosilicon molecule has ever been isolated, such molecules are implied by the presence of silica as supporting tissue in sponges, diatoms, radiolarians, and some higher plants. Such biostructures imply a robust silicate biochemistry, which is largely unknown at present. Progress has been made primarily in understanding the proteins, called silacateins, which are responsible for biouptake of aqueous silicate and production of ordered silica nanostructures [6, 7].

The seminal experiments in identifying interactions between the organic world of carbohydrates and the inorganic world of silicates were reported by the groups of Kinrade and of Klüfers in 1999 [8, 9–11]. Kinrade [8] reported that aqueous sodium silicate reacts with certain glycitols (linear polyols with a single hydroxy group on each carbon, e.g., $HOCH_2CHOHCHOHCH_2OH$, as well as glyconic acids (open chain sugars in which the aldehyde has been oxidized to the carboxylic acid, e.g., $HOCH_2CHOHCHOHCO_2H$. They used ^{29}Si NMR spectroscopy to confirm that the glycitols and glyconic acids are complexed with pentacovalent (negatively charged) silicon. Klüfers [10] inferred that the complexes between silicate and glycitols comprise five-membered diolato rings. They confirmed the five-membered structures with X-ray analysis of the glycitols of mannose, xylose, and threose, [11] which proved to form 3:1 hexacoordinated diolato complexes with silicate.

In contrast to glycitols and glyconic acids, sugars (glycoses) exist as cyclic structures. Thus the linear form, e.g., $HOCH_2CHOHCHOHCHO$, is converted to a ring form, which can have either five members (furanose) or six members (pyranose), as illustrated in **1** and **2** for ribose.

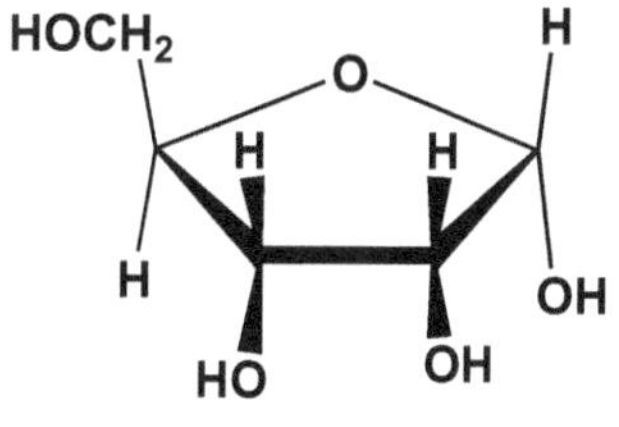

HOCH₂ O H H H H OH HO OH

1

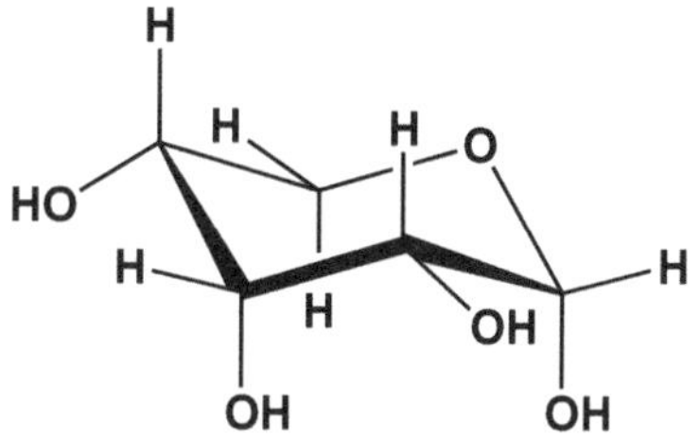

H H H O HO H H H OH OH OH

2

Four-carbon sugars can exist only as furanoses, and smaller sugars cannot exist in the cyclic forms. Sugars with five or more carbons exist in both furanose and pyranose forms. Like the glycitols and glyconic acids, sugars have strings of hydroxy groups, which conceivably could complex with silicate. Given the stereochemical restrictions that Kinrade and Klüfers had found for the glycitols and glyconic acids, it was likely that sugars would have similar or stronger restrictions, as the rings in which they exist are conformationally more restricted. In an early report, Kinrade and co-workers [12] found that ribose forms a complex with silicate.

Lambert and co-workers reported the first comprehensive examination of sugar silicates in 2004 [13]. They examined all four aldopentoses, seven of the eight aldohexoses, all four ketohexoses, several disacchraides, and the 1-*O*-methyl glycosides of several sugars. Of these, the monosacchraides ribose, xylose, lyxose, talose, psicose, fructose, sorbose, and tagotose, and the disaccharides lactulose, maltulose, and palatinose successfully formed sugar silicate complexes, as signified by observation of ^{29}Si signals in the pentacoordinate region. Notable failures include the monosacchrides arabinose, glucose, mannose, and galactose, the disaccharides sucrose and turanose, and all the 1-*O*-methyl glycosides. For a positive result, three structural factors had to be operative: (1) The anomeric oxygen must be free, that is without *O*-methyl glycosidation. The anomeric hydroxy hydrogen is the most acidic in the molecule and probably initiates the reaction. (2) The carbon adjacent to the anomeric carbon must bear a hydroxy group *cis* to the anomeric hydroxyl. By a chelating effect, silicate complexes with two of the sugar hydroxyls, and *cis*-diols are the most amenable. (3) These elements must reside in a furanose, not a pyranose, ring. Apparently puckering of the six-membered ring sterically inhibits formation of the five-membered oxalato ring. The resulting complex would take the form **3** for D-ribose.

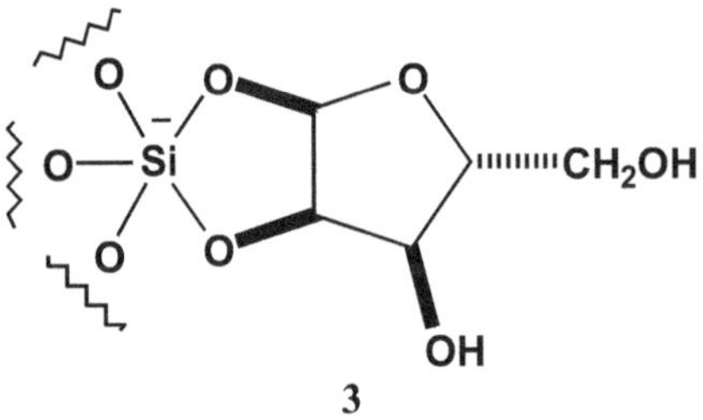

3

The common sugars glucose and mannose lack significant amounts of the key furanose form. All five disaccharides contain furanose rings, but in sucrose the two anomeric carbons are bonded through a glycosidic linkage, so the key anomeric hydroxyl is blocked. In tagotose the 2 hydroxyl is linked to the second sugar ring, so the molecule lacks a cis diol structure that includes a furanoid anomeric hydroxyl. Since such a structure is available in lactulose, maltulose, and palatinose, those three disaccharides form sugar silicates.

The final element of the structure of the complexes involved the question of stoichiometry, which for structure **3** could be a 1:1 combination of sugar and silicate in which the remaining three linkages to silicon are hydroxyls, or a 2:1 sugar/silicate combination in which one of the linkages is hydroxyl and the other two are a second sugar. Electrospray mass spectral examination of the complexes indicated that all are 2:1 complexes, such as **4** for ribose [13].

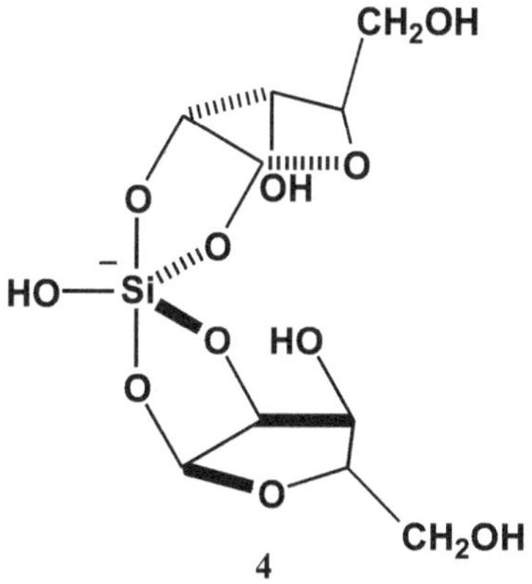

4

Such structures can exist as several stereoisomers. The illustrated structure for ribose silicate has the two ribose rings *syn* to each other and pointing away from the remaining silicate hydroxyl. The two groups could by *syn* to each and pointing towards the silicate hydroxyl, or they could be anti to each other with one pointing towards and the other away from the silicate hydroxyl. There also could be regioisomers, as the two sugar rings could be arranged with a different relationship of the C1 and C2 bonds, for example. The ^{29}Si spectra of sugar silicates exhibit multiple peaks (usually three to five), representing the multiple isomers. Unfortunately, these solutions never yielded crystalline materials from which X-ray structures could be obtained. In 2005, Klüfers et al. obtained such structures in nonaqueous media, although with phenyl in place of the silicate hydroxyl, confirming all aspects of the structures, including furanose rings, *cis*-diol coordination, and 2:1 stoichiometry [14]. These experiments by Kinrade, Lambert, and Klüfers established a rich link between carbohydrates and silicates.

These reactions take place in highly basically medium, typically pH 10.5. Sugars are famously unstable in base, isomerizing from aldehydes to ketones, undergoing aldol condensations, and breaking apart in reverse aldols. The silicate complexes, however, are reasonably stable for hours to days under these highly basic conditions, without isomerization or decomposition. Silicate formation, moreover, is almost instantaneous even at room temperature. This distinction between sugars and sugar silicates under basic conditions has ramifications in the field of prebiotic chemistry. On the one hand, silicate formation could be a mechanism for sequestering sugars under extreme conditions, for example as found on comets lacking an atmosphere. The higher stability and lower volatility of sugar silicates could allow them to abide in space until they arrive at a planet with an atmosphere. Such conjectures are highly speculative. On the other hand, silicate formation could provide the stabilization mechanism that Reid and Orgel stated was necessary before the formose reaction could be considered practical for the prebiotic synthesis of oligosaccharides (“some method of stabilizing the sugars is essential”) [4].

Lambert and co-workers reported the realization of this possibility in what they termed the bottom-up synthesis of sugar silicates (in distinction to the direct reaction between, or top-down synthesis of, sugars with silicate) [15]. They compared various formose reactions either with sodium hydroxide as catalyst (classic formose conditions) or with sodium silicate as catalyst. They found that formaldehyde (C1) alone produces low yields under both conditions, as it lacks the alpha hydrogen to initiate the condensation. They studied the reactions of glycolaldehyde (C2) and glyceraldehyde (C3), either alone or mixed with the other or with C1. These sugars exist in the straight chain and other forms, but they do not have enough carbons to form a furanose ring. In the case of C3 alone, the reaction occurs with either catalyst to form C6 oligomers within seconds at room temperature. Under standard formose conditions, however, the oligomers decompose quickly, so that within 12 h little sugar product remains. When the same reaction is carried out with C3 in the presence of sodium silicate as catalyst, solely C6 products are formed, remaining almost unchanged for 12 h or more. Similar results occur with C2 alone. The reaction of C2 and C3 together is more complex, because it can proceed from simple dimerization to give C4 and C6, but also by the cross reaction to give C5. It appears that C5 products are the most abundant, and, again, the products are unstable in the presence of sodium hydroxide but robust when the catalyst is sodium silicate.

The possibility that silicate mediation can stabilize the products of the formose reaction may revive its role as the mode for prebiotic synthesis of sugars. Silicate minerals are widely available, although the basic conditions are less available in Nature. On earth, such conditions occur naturally under conditions of extreme evaporation and concentration, as occurs in the Dead Sea, the Great Salt Lake, and the lakes of the Atacama Desert of Chile [16]. A similar rationale had been proposed with borate minerals [17, 18]. Although borates may be kinetically more effective than silicates, their much lower availability makes them a less likely stabilizing agent under prebiotic conditions [19]. Moreover, Grew et al. have pointed out that the development of borate minerals in the Earth's crust may have occurred too late to be useful for prebiotic processes: "concentrations of B either on land or in the sea sufficient to play a role in ribose stabilization are thus unlikely" [20].

Addressing a different issue, Vázquez-Mayagoitia et al. carried out *ab initio* calculations at the B3LYP level on silicate complexes of the C5 sugars arabinose, lyxose, ribose, and xylose[21]. They studied five different stereochemical versions for each sugar, such as **4**. They found that the ribose silicates were more stable than the other sugar silicates, "to the extent that the least stable of these is even more stable than the most stable stereoisomer of the other 2:1 sugar-silicate complexes." They suggested that formose reactions in the presence of sodium silicate should form ribose products preferentially over the other C5 sugars.

2.4 Summary

In summary, polyhydroxy compounds readily form complexes with sodium silicate at room temperature. Stereochemical requirements limit the reaction to sugars that exist in the furanoid form, have an unsubstituted anomeric hydroxy group, and have a *syn* hydroxyl adjacent to the anomeric hydroxyl. Ribose is one such sugar with the appropriate stereochemistry. Once in the silicate form, sugars are far more stable under basic conditions. This observation suggests a sequestering mechanism for sugars under extreme conditions. It also suggests a mechanism for stabilizing sugars as they are formed during the formose reaction. This aldol reaction of small sugars had been discarded as a process for the prebiotic synthesis of sugars because of their rapid decomposition. Indeed, glyceraldehyde (C3) dimerizes to a stable solution of C6 sugar silicates in the presence of sodium silicate but dimerizes to an unstable solution in the presence of sodium hydroxide. Similarly in the presence of sodium silicate, glycolaldehyde (C2) leads to a stable solution of C4, and a mixture of C2 and C3 leads to a stable solution containing C5 as the primary product. Based on calculations ribose forms the most stable sugar silicate. Stabilization also could be realized with borate salts, but their role in the prebiotic environment has been questioned on geochemical grounds.

References

1. Miller SL (1953) Science 117:528
2. Oró J, Kimball AP (1961) Arch Biochem Biophys 94:217–227
3. Gabel NW, Ponnamperuma C (1967) Nature 216:453–455
4. Breslow R (1959) Tetrahedron Lett 1:22
5. Reid C, Orgel LE (1967) Nature 216:455
6. Cha J, Shimizu K, Zhou Y, Christiansen SC, Chmelka BF, Stucky GD, Morse DE (1999) Proc Natl Acad Sci USA 96:361–365
7. Kröger N, Lorenz S, Brunner E, Sumper M (2002) Science 298:584–586
8. Kinrade SD, Del Nin JW, Schach AS, Sloan TA, Wilson KL, Knight CTG (1999) Science 285:1542–1545
9. Kinrade SD, Hamilton RJ, Schach AS, Knight CTB (2001) J Chem Soc Dalton Trans 961–963
10. Benner K, Klüfers P, Schuhmacher J (1999) Z Anorg Allg Chem 625:541–543
11. Benner K, Klüfers P, Vogt M (2003) Angew Chem Int Ed 42:1058–1062
12. Kinrade SD, Deguns EW, Gillson A-ME, Knight CTG (2003) J Chem Soc Dalton Trans 3713–3716
13. Lambert JB, Lu G, Singer SR, Kolb VM (2004) J Am Chem Soc 126:9611–9625
14. Kästele X, Klüfers P, Kopp F, Schuhmacher J, Vogt M (2005) Chem Eur J 11:6326–6346
15. Lambert JB, Gurusamy-Thangavelu SA, Ma K (2010) Science 327:984–986
16. Stumm W, Morgan JJ (1996) Aquatic Chemistry 3rd edn. Wiley-Interscience, New York
17. Ricardo A, Carrigan MA, Olcott AN, Benner SA (2004) Science 303:196
18. Kim H-J, Benner SA (2010) Science 329:902–a
19. Lambert JB, Gurusamy-Thangavelu SA, Ma K (2010) Science 329:902–b
20. Grew ES, Bada JL, Hazen RM (2011) Orig Life Evol Biosph 41:307–316
21. Vázquez-Mayagoitia Á, Horton SR, Sumpter BG Šponer J, Šponer JE, Feuntes-Cabrera M (2011) Astrobiology 11:115–121

Chapter 3
Protease-Mediated Hydrolysis and Condensation of Tetra- and Trialkoxysilanes

Mark B. Frampton and Paul M. Zelisko

3.1 Introduction

Many marine organisms and terrestrial plants have evolved molecular machinery to process silicic acid or sodium silicate from their local environments to form elaborate shells, frustules, [1–3] spicules, [4] or other structural elements [5–7]. Organisms typically process silica under ambient temperatures, standard pressure, and near neutral pH, where the external concentration of silica is generally in the micromolar range. The ability of these organisms to form elaborate structures only serves to highlight the important mechanisms that these organisms have evolved to process silicon-based compounds.

3.1.1 Biosilica Synthesis

The synthesis of silica-derived materials usually requires extremes of temperature, pressure and pH, while diatoms and marine sponges have evolved biochemical processes to direct the templating of inorganic silica. In diatoms there have been two major silica precipitating proteins that have been identified, the silaffins and silacidins. The silaffins (silaffin-$1A_1$, $1A_2$, 1B and 2) so named for their *sil*ica *affin*ity, have been studied at the molecular level and their primary amino acid sequence determined [1, 2]. The amino acid sequence of silaffin−$1A_1$ and $1A_2$ indicate a high serine and lysine content. Many of the serine residues are phosphorylated and the lysine residues are almost universally found with post-translational modifications (Fig. 3.1) [2, 8].

P. M. Zelisko (✉) · M. B. Frampton
Department of Chemistry, Centre for Biotechnology, Brock University, 500 Glenridge Avenue, St. Catharines, ON L2S 3A1, Canada
Tel.: +1-905-688-5550
e-mail: pzelisko@brocku.ca

P. M. Zelisko (ed.), *Bio-Inspired Silicon-Based Materials,* Advances in Silicon Science 5,
DOI 10.1007/978-94-017-9439-8_3

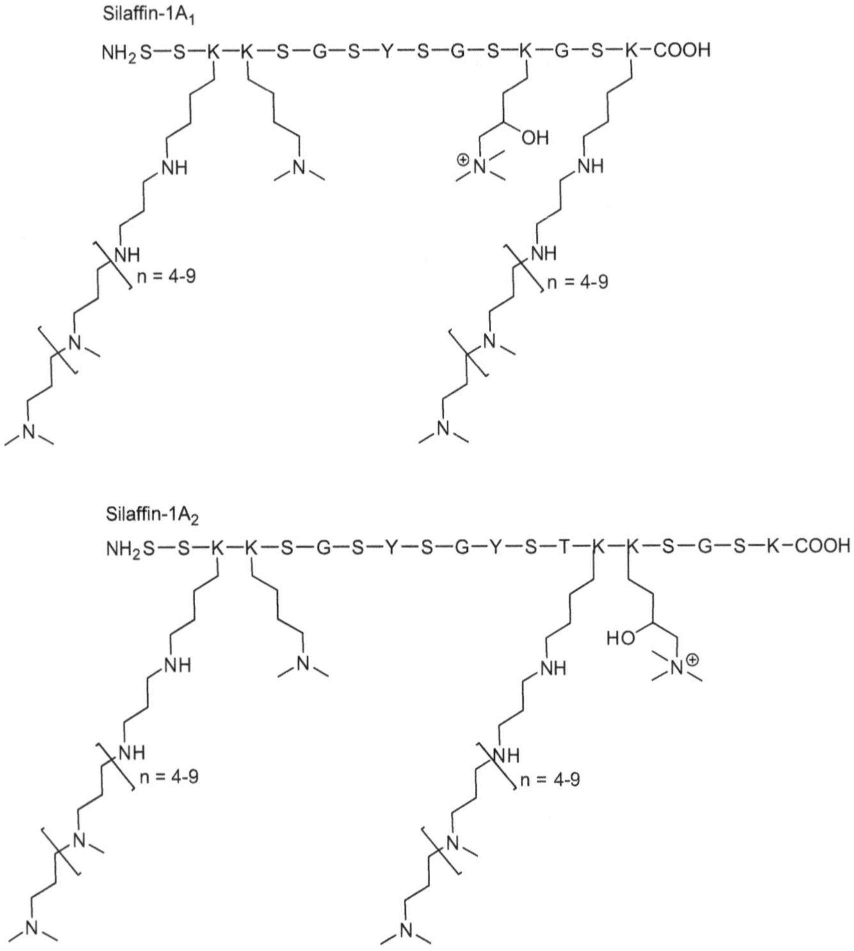

Silaffin-1A$_1$

NH$_2$S—S—K—K—S—G—S—Y—S—G—S—K—G—S—K—COOH

Silaffin-1A$_2$

NH$_2$S—S—K—K—S—G—S—Y—S—G—Y—S—T—K—K—S—G—S—K—COOH

Fig. 3.1 The chemical structures of silaffin-1A$_1$ and silaffin-1A$_2$. The post-translationally modified lysine residues contribute to the structure directing activity of the silaffins [2, 9]

The silica precipitating action of silaffin-1A is dependent on these post-translationally modified lysine residues [2]. Silica precipitation assays using silaffin preparations with a metastable silicic acid solution produced silica within seconds [1]. The silaffins co-precipitated with silica suggesting that they act in a structure directing role.

The silacidins on the other hand are rich in aspartate/glutamate and phosphorylated serine residues that make them anionic at neutral pH [3]. These peptides produced silica, when combined with polyamines, in a concentration dependent manner and it has been hypothesized that these proteins are the biological anion required to direct that precipitation of silica *in vivo*.

Long chain polyamines (LCPs), isolated from diatoms function in tandem with silaffins and silacidins to direct the synthesis of bio-silica. The addition of LCPs to silicic acid solutions induces silica formation via an electrostatic interaction between the amines and silanols of silicic acid [9]. Sumper proposed a phase separation model to account for the action of LCPs in buffered solution [10, 11].

The other major group of silica precipitating organisms are marine sponges which possess three related enzymes, silicatein α, β and γ, which are required for spicule formation [4, 12, 13]. Silicateins are structurally homologous to the caspase and papain families of cysteine proteases but instead of a cysteine residue, a serine is used as the active nucleophile making these enzymes homologous to the serine family of hydrolases [4]. Silicatein has been shown to precipitate amorphous silica from buffered solutions using tetraethoxysilane as a silica precursor [13].

Enzymes that have not been traditionally associated with processing silica or silica precursors have been used to perform or enhance the rate of silica chemistry; most notably, trypsin, [14–16] α-chymotrypsin, [16] lysozyme, [17, 18] lipases, [19, 20] bovine serum albumin [17] and carbonic anhydrase [21].

3.2 Enzyme-Mediated Hydrolysis and Condensation of Alkoxysilanes

Building upon early work by the Morse group, we embarked on a research avenue focusing on the interaction between enzymes and silicon based substrates with a focus on finding enzymes that had a capacity for catalytic activity at silicon . Trypsin and α-chymotrypsin were originally thought to be suitable candidates as catalysts that could be used under mild conditions (i.e., near neutral pH, low ionic strength, and low temperature and pressure) [22]. When either of the enzymes was charged with TEOS, amorphous monolithic silica gels were produced. The hydrolysis of TEOS by trypsin was generally slow, requiring 24–48 h. α-Chymotrypsin also showed the capacity to produce amorphous silica that was macroscopically indistinguishable from that produced by other modes of catalysis. After approximately 24 h a clear and colorless, viscous, monophasic sol appeared. Prolonged ageing at room temperature resulted in the formation of a hard silica monolith. To ensure that silica formation was the result of an enzymatic process the active site of trypsin and α-chymotrypsin were each blocked by a soybean inhibitor [23]. Prolonged reaction times did not promote silica formation with the inhibited enzymes; in separate experiments, the soybean inhibitor did not demonstrate any catalytic activity towards TEOS. Solid-state ^{29}Si NMR spectra of enzyme-produced silica gels suggested that catalyst choice had only a marginal effect on the extent of siloxane condensation (Fig. 3.2).

We expanded the scope of our investigation to include other enzymes such as pepsin (carboxypeptidase), lipase from *Candida rugosa* (serine hydrolase), human serum albumin, bromelain (cysteine protease from pineapple) and papain (cysteine protease from papaya). In addition, the number of alkoxysilanes screened was expanded to include methyl-, ethyl-, allyl- and phenyl-trimethoxysilane, as well as a select few bis(triethoxysilyl) alkanes (Fig. 3.3).

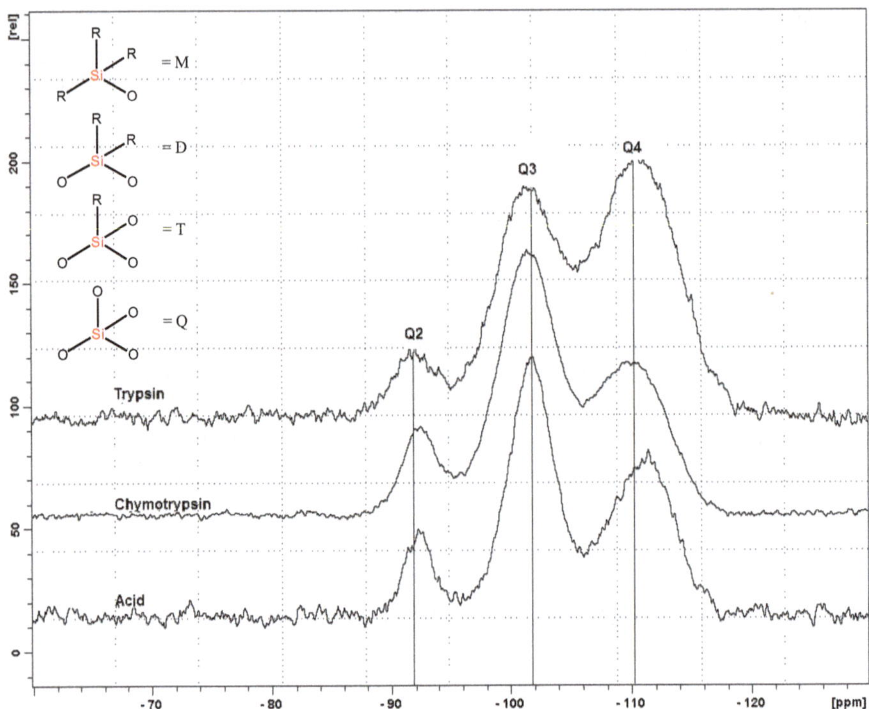

Fig. 3.2 Solid-state ^{29}Si NMR spectra of three amorphous silica gels produced using trypsin (*top*), α-chymotrypsin (*middle*) and hydrochloric acid (*bottom*) [22]

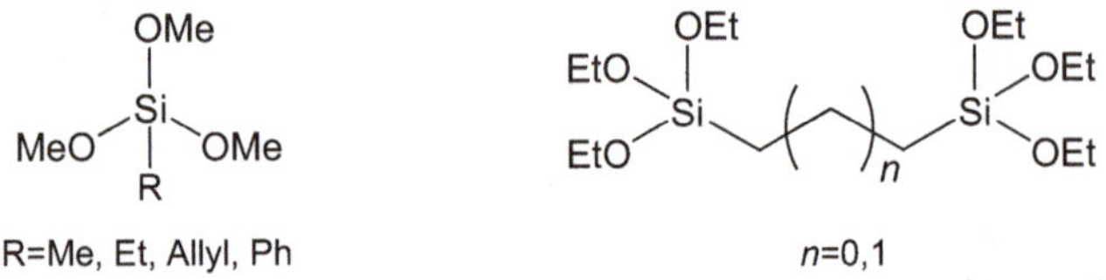

R=Me, Et, Allyl, Ph

n=0,1

Fig. 3.3 The alkoxysilanes that were subjected to enzymatic hydrolysis by enzymes not naturally evolved to process silicon-based substrates

In addition to trypsin and α-chymotrypsin, the combinations of pepsin or bromelain with TEOS lead to the formation of silica monoliths. The other enzymes did not appear to hydrolyze TEOS as well and produced only minute amounts of precipitated silica. When these same seven enzymes were charged with the hydrolysis of phenyltrimethoxysilane (PTMS), only trypsin, α-chymotrypsin, and pepsin produced organically modified silica.

Attempts at using enzymes to process bis(trialkoxysilyl)alkanes were not fruitful. In fact analysis by ^{29}Si NMR could not confirm even a single hydrolysis event. The lack of enzyme-mediated chemistry has been attributed to phase separation between the hydrophobic bis(trialkoxysilyl)alkane and the aqueous enzyme solution.

3.3 Hydrolysis and Condensation of Organically Modified Alkoxysilanes

Trypsin, α-chymotrypsin, and pepsin were chosen as candidate enzymes with which to study the hydrolysis of trialkoxysilanes. Methyltrimethoxysilane (MTMS), ethyltrimethoxysilane (ETMS), allyltrimethoxysilane (ATMS) and PTMS were selected as the candidate alkoxysilanes. To study the reaction(s) of the enzymes with the alkoxysilanes ^{29}Si NMR was used as an invaluable technique for following the hydrolysis and condensation of the alkoxysilanes given its quantitative nature and the ease with which it allows for the identification of each unique ^{29}Si environment.

The first approach studied the interaction between trypsin and MTMS. While visual evidence supported the expedient hydrolysis of MTMS, enzyme-free reactions also hydrolyzed within 5–10 min. The hydrolysis of MTMS could not be followed using our ^{29}Si NMR method as hydrolysis was on a time scale that was too quick for the NMR experiments; each spectrum requiring 60 min to complete. As such it was not possible to quantify any difference between the background and trypsin-mediated rates of hydrolysis.

Visual inspection of the enzyme-free hydrolysis of ETMS suggested that hydrolysis did not occur as the reaction mixtures always remained biphasic with no evidence for the precipitation of solids. ^{29}Si NMR experiments refuted this observation and provided data suggesting that uncatalyzed hydrolysis was in fact occurring. While the contents of the NMR reaction remained biphasic, a resonance at −37 ppm was observed after the first hour of incubation and was subsequently identified as ethylsilanetriol [24]. To confirm that no other hydrolysis or condensation products were remaining in the supernatant, it was analyzed and the data showed that it was almost exclusively composed of ETMS with only a small amount of the ethylsilanetriol present.

The combination of trypsin in 3:1 D_2O:H_2O and ETMS led to the hydrolysis and condensation of ETMS, occurring within the first hour (See Table 3.1 for a complete list of all identified products). Within 16 h there was little spectroscopic evidence for any remaining ETMS. Ethylsilanetriol reached a maximum concentration by 10 h, and persisted until 80 h. Dimerization products (−46.8 ppm) and some higher order oligomers (−56.8 ppm) started to appear when ethylsilanetriol reached a maximum. Additionally resonances appeared at −50.2 ppm after the first hour but were completely absent after 24 h. These resonances were not seen in any experiments with the other alkoxysilanes. A comparison of these resonances with those provided by Sugahara et al suggested that the resonance at −50.2 ppm was 1,3-diethyl-1,1,3,3-tetramethoxydisiloxane [25]. However, without visual evidence for the formation of ethyldimethoxysilanol it is difficult to offer a mechanistic explanation as to how this particular dimer arises.

Visual inspection of enzyme-free control reactions performed with ATMS suggested that spontaneous hydrolysis was not appreciable with this particular alkoxysilane. NMR-scale reactions, however, demonstrated that in fact hydrolysis and condensation were occurring in the absence of the enzyme catalyst, albeit to a very small extent. It is worth noting that the parent peak for ATMS, normally located at −48.0 ppm was not visible in any of the ^{29}Si NMR spectra that were acquired suggesting a rapid phase separation, which could contribute to the apparent slow rate of hydrolysis.

Table 3.1 Assignment of ^{29}Si resonances for the organically modified trimethoxysilanes

R	Peak assignment	δ (ppm)	Structural feature
Phenyl	T^3_0	−51	RSi(OH)3
	T^0_0	−55	RSi(OMe)3
	$T^2_1T^2_1$	−61.1	$R(OH)_2\underline{Si}$-O-$\underline{Si}(OH)_2R$
	$T^2_1T^1_2T^2_1$	−60.9	$R(OH)_2\underline{Si}$-O-SiR(OH)-O-$\underline{Si}(OH)_2R$
	$T^2_1T^1_2T^2_1$	−70.5	$R(OH)_2Si$-O-$\underline{Si}R(OH)$-O-$Si(OH)_2R$
Ethyl	T^3_0	−37.2	$R\underline{Si}(OH)_3$
	T^0_0	−42.2	$R\underline{Si}(OMe)_3$
	$T^2_1T^2_1$	−46.8	$R(OH)_2\underline{Si}$-O-$\underline{Si}(OH)_2R$
	$T^0_1T^0_1$	−50.2	$R(OMe)_2\underline{Si}$-O-$\underline{Si}(OMe)_2R$
	$T^2_1T^1_2T^2_1$	−56.8	$R(OH)_2Si$-O-$\underline{Si}R(OH)$-O-$Si(OH)_2R$
Allyl	T^3_0	−43.3	$R\underline{Si}(OH)_3$
	T^0_0	−48.0	$R\underline{Si}(OMe)_3$
	$T^2_1T^2_1$	−53.0	$R(OH)_2\underline{Si}$-O-$\underline{Si}(OH)_2R$
	$T^2_1T^1_2T^2_1$	−63.1	$R(OH)_2Si$-O-$\underline{Si}R(OH)$-O-$Si(OH)_2R$

On the other hand, the trypsin-mediated hydrolysis of ATMS proceeded rapidly, with all of the ATMS being consumed within the first 10 h. Unlike the enzyme-free control experiment in which the ^{29}Si resonance for ATMS was not visible, it was clearly seen in the spectra for the trypsin-mediated processes. The only hydrolysis product that was evident was allylsilanetriol (−43.0 ppm). Dimerization products (−53.0 ppm) were visible after 7 h. These disiloxanes were difficult to identify, but given the lack of spectroscopic evidence for the first or second hydrolysis products, it is likely that the resonance can be attributed to the fully hydrated disiloxane, 1,3-diallyl-1,1,3,3-tetrahydroxydisiloxane. Higher order oligomers were not spectroscopically determinable until after 22 h and resonated at −63.1 ppm.

Under enzyme-free conditions, there was little evidence to suggest that any hydrolysis of PTMS occurred; the only visible resonance was located at −56.0 ppm which corresponded to PTMS. An additional resonance appeared at −51.5 ppm during the first hour of spectral acquisition when trypsin was included in the reaction mixture, corresponding to phenylsilanetriol. The identities of these resonances were derived from those provided for phentriethoxysilane [26] and methyltrimethoxysilane [25, 27]. As with the previous experiments in this series, the first and second hydrolysis products were not detected.

After 13 h of spectral acquisition, a third and fourth resonance appeared at −61.1 ppm and −70.5 ppm as disiloxane and oligomer formation became favoured processes. The peak at −61.1 ppm was assigned as the fully hydroxylated disiloxane and the peak at −70.5 ppm as the fully hydroxylated trisiloxane. However, line broadening may have obscured other resonances. The presence of the 1,3,5-triphenyl-1,1,3,5,5-pentahydroxytrisiloxane oligomer was further confirmed by the growth of a smaller peak at −60.9 ppm which we believe to be the end groups of that oligomer.

3.4 Active Site Considerations

Enzymes have evolved to accept substrates with particular molecular geometries and functional groups. Part of this specificity can be attributed to the local environments of the secondary binding pockets located within or proximal to the active site. There are many evolutionary conserved residues shared between trypsin and α-chymotrypsin, but it is their differences that permit the binding of different substrates. The *S1* binding pockets of the serine proteases are formed by residues 189–195, 214–220, and 225–228 [28] of which four of these residues are not conserved between trypsin and α-chymotrypsin; residues 189, 192, 217 and 219. For example Ser_{190} is differently oriented within each enzyme; pointing into the *S1* binding pocket of trypsin but in α-chymotrypsin it points away from the *S1* binding pocket where it hydrogen bonds to Thr_{138} [28]. The bottom of the *S1* site in trypsin and α-chymotrypsin show an additional dissimilarity; trypsin contains an Asp_{189} which is anionic at neutral pH and stabilizes the positive charge of the side chains of lysine and arginine residues [29]. The *S1* binding pocket of α-chymotrypsin, on the other hand, does not contain this anionic group allowing it to accept hydrophobic amino acid side chains [30].

Pepsin, on the other hand, is a carboxypeptidase and contains two binding pockets adapted for aromatic rings that are located proximally to the catalytic carboxylic acid residues, Asp_{215} and Asp_{32}. These residues serve to polarize a single molecule of water making it more nucleophilic towards the peptide bond to be cleaved.

The hydrolysis of PTMS by these three enzymes was followed by ^{29}Si NMR spectroscopy. While the conditions that were employed were not considered optimal for any of these enzymes, it was important to minimize any background hydrolysis rate, known to be slowest at neutral pH, [5] and to treat each enzyme with identical processing parameters. The hypothesis was that there was a correlation between the architecture of the *S1* binding pocket and the rate of hydrolysis of PTMS.

The integration values from each ^{29}Si NMR spectrum can be used to follow the hydrolysis of PTMS. A plot of $\ln[^{29}Si]$ versus time results in a straight line, the slope of which is the observed pseudo first order rate constant [23, 31]. The pseudo first order rate constants derived from these experiments are summarized in Table 3.2 and Fig. 3.4.

Increasing the enzyme concentration was followed by an increase in the rate constant (Fig. 3.4) [32]. Trypsin was the least proficient at hydrolyzing PTMS. Furthermore, trypsin also showed the smallest changes in the rate constant as the amount of the enzyme was increased. On the other hand pepsin displayed the highest rates of hydrolysis, and like trypsin, exhibited only small changes in the rate constant as the concentration was increased. If the line representing pepsin was to be extrapolated to lower concentrations it would not pass through the origin. Unfortunately we cannot make any comment on the behaviour of the enzyme at those concentrations as a result of difficulties in obtaining reliable ^{29}Si NMR data. The rate of hydrolysis of PTMS by α-chymotrypsin at lower concentrations was intermediate compared to trypsin and pepsin, but as the amount α-chymotrypsin was increased, the associated change in the rate constant was the largest of the three enzymes.

Table 3.2 Summary of the pseudo first order rate constants for the enzyme-mediated hydrolysis of PTMS using different concentrations of enzyme. Rate constants (h^{-1}) are reported as the average of at least triplicate trails±standard deviation. Table adapted from Reference 32

Enzyme	5 mg/mL	7.5 mg/mL	10.0 mg/mL	12.5 mg/mL	15.0 mg/mL	20.0 mg/mL
Trypsin	–	0.02±0.0035	0.04±0.1322	14±0.0750	0.22±0.0002	–
α-Chymotrypsin	0.11±0.0050	0.44±0.1050	0.54±0.2557	0.83±0.3310	1.10±0.2152	–
Pepsin	0.54±0.2190	–	0.67±.1070	–	0.86±0.0000	0.99±0.0948

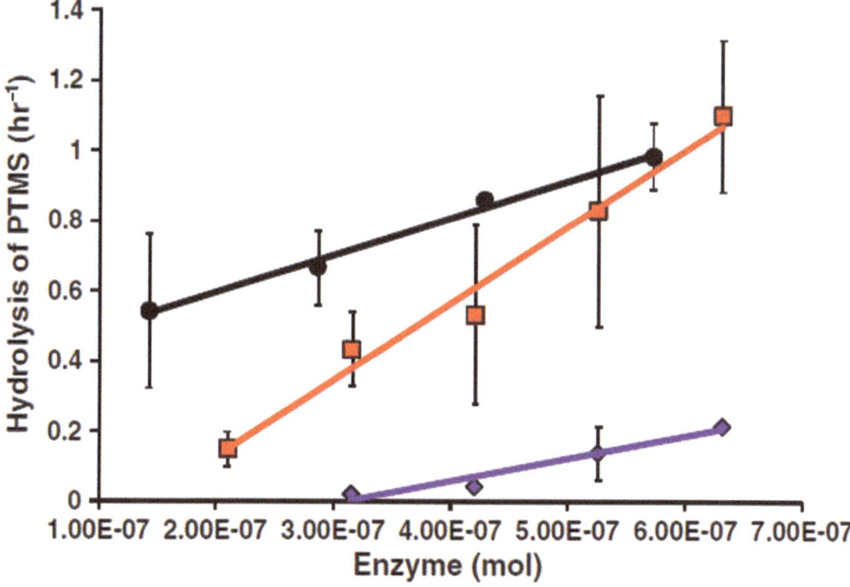

Fig. 3.4 The hydrolysis of PTMS. The pseudo first order rate constant increases as the enzyme concentration is increased. Legend: Pepsin-*black circles*, α-chymotrypsin-*red squares*, trypsin-*purple diamonds*. Error bars indicate the standard deviation of triplicate trials [32]

The observed enhanced rate of hydrolysis by pepsin can be explained in two ways. Catalysis by pepsin is carried out by an aspartic acid dyad. One of the aspartic acid residues exists as the acid form while the other exists as the carboxylate. Working in tandem these two amino acids polarize a water molecule making it an active nucleophile. An acyl enzyme intermediate is not formed and if a similar mode of action is in place here then the absence of a covalently bound PTMS in the active site may serve to facilitate hydrolysis. Like pepsin, α-chymotrypsin favorably binds the aromatic side chains of several amino acids. The observed differences may potentially be attributed to the formation of an acyl-silane intermediate between α-chymotrypsin and PTMS in a manner similar to that proposed for the silicatein-mediated hydrolysis of TEOS. The enhanced catalytic activity of α-chymotrypsin over trypsin may be attributed to the binding of PTMS in the *S1* pocket; the *S1* pocket of trypsin has not evolved to accept a hydrophobic phenyl ring.

3.5 Conclusions

Several enzymes have been shown to be catalytically active towards some alkoxysilanes. ^{29}Si NMR spectroscopy was employed to elucidate the intermediates along the reaction pathway of the hydrolysis of some of these alkoxysilanes. While not all enzymes have a capacity for performing such chemistry, those that can must have an available and catalytically competent active site, but most also possess favourable secondary interactions counter within the active site. Enzymes possessing a preference for hydrophobic aromatic amino acids have a greater capacity for hydrolyzing PTMS and as such show enhanced hydrolysis rates. While a complete and accurate reaction mechanism is debated, theoretical calculations suggest that the first step of the mechanism proceeds through serine addition to the alkoxysilane.

Acknowledgements The authors would like to thank Razvan Simionescu (Brock University) for assistance in acquiring NMR spectra. Funding for these projects was provided by Brock University, Natural Science and Engineering Research Council (NSERC), and the Ontario Partnership for Industrialization and Commercialization (OPIC). MBF was supported through graduate scholarships from the Ontario Scholarship (OGS), the Ontario Graduate Scholarship in Science and Technology (OGSST) and the Queen Elizabeth II Graduate Scholarship in Science and Technology (QEII-GSST) programs.

References

1. Kroger N, Deutzmann R, Sumper M (1999) Science 286:1129–1132
2. Kroger N, Deutzmann R, Sumper M (2001) J Biol Chem 276:26066–26070
3. Wenzl S, Hett R, Richthammer P, Sumper M (2008) Angew Chem Int Ed 7:1729–1732
4. Shimizu K, Cha JN, Stucky GD, Morse DE (1998) Proc Natl Acad Sci U S A 95:6234–6238
5. Iler RK (1979) The chemistry of silica, solubility, polymerization, colloid and surface properties, and biochemistry. Wiley, New York
6. Perry CC, Keeling-Tucker T (1998) Chem Commun 23:2587–2588
7. Perry CC, Keeling-Tucker T (2003) Colloid Polym Synth 81:652–664
8. Kroger N, Lorenz S, Brunner E, Sumper M (2002) Science 298:584–586
9. Kroger N, Deutzmann R, Bergsdorf C, Sumper M (2000) Proc Natl Acad Sci U S A 97:14133–14138
10. Sumper M (2002) Science 295:2430–2433
11. Sumper M (2004) Angew Chem Int Ed 43:2251–2254
12. Zhou Y, Shimizu K, Cha JN, Stucky GD, Morse DE (1999) Angew Chem Int Ed 38:779–782
13. Cha JN, Shimizu K, Zhou Y, Christiansen SC, Chmelka BF, Stucky GD, Morse DE (1999) Proc Natl Acad Sci U S A 96:361–365
14. Bassindale AR, Branstadt KF, Lane TH, Taylor PG (2003) J Inorg Biochem 96:401–406
15. Zelisko PM, Dudding T, Arnelien KR, Stanisic H (2010) In: Clarson SJ, Owen MJ, Smith SD, Van Dyke ME (eds.) Advance of silicones and silicone-modified materials. Chap. 5, pp 47–57
16. Frampton M, Vawda A, Fletcher J, Zelisko PM (2008) Chem Commun 43:5544–5546
17. Coradin T, Coupé A, Livages J (2003) Colloid Surf B Biointerfaces 29:189–196
18. Abbate V, Bassindale AR, Brandstadt KF, Lawson R, Taylor PG (2010) Dalton Trans 39:9361–9368
19. Buisson P, El Rassy H, Maury S, Pierre AC (2003) J Sol-Gel Sci Technol 27:373–379
20. Pierre AC, Buisson P (2006) J Sol-Gel Sci Technol 38:63–72
21. Favre N, Ahmad Y, Pierre AC (2011) J Sol-Gel Sci Technol 58:442–451
22. Frampton M, Vawda A, Fletcher J, Zelisko PM (2008) Chem Commun 43:5544–5546
23. Frampton MB, Simionescu R, Zelisko PM (2009) Silicon 1:47–56
24. Hook RJ (1996) J Non-Cryst Solids 195:1–16
25. Sugahara Y, Okada S, Sato S, Kuroda K, Kato C (1994) J Non-Cryst Solids 167:21–28
26. Kuniyoshi M, Takahashi M, Tokuda Y, Yoko T (2004) J Sol-Gel Sci Technol 39:175–183

27. Sugahara Y, Inoue T, Kuroda K (1997) J Mater Chem 7:53–59
28. Hung SH, Hedstrom L (1998) Prot Eng 11:669–673
29. Voet D, Voet JG (1990) Biochemistry, 3rd edn. Wiley, New York
30. Berg JM, Tymoczko J, Stryer J (2002) Biochemistry, 5th edn. W.H. Freeman and Company, New York
31. Frampton MB, Simionescu R, Dudding T, Zelisko PM (2010) J Mol Cat B Enz 66:105–112
32. Frampton MB, Zelisko PM (2012) Silicon 4:51–56

Chapter 4
Bioinspired Silica for Enzyme Immobilisation: A Comparison with Traditional Methods

Claire Forsyth and Siddharth V. Patwardhan

4.1 Introduction to Enzymes

In recent years, it has become necessary to investigate the feasibility of more environmentally friendly or '*green*' process routes. This is largely due to increasingly stringent environmental regulations, environmental concerns, and drives for greater efficiency [1]. Compared with traditional routes, green routes typically involve lower temperatures, therefore reducing energy costs, and minimise the use of any toxic materials or the production of harmful waste. These can greatly reduce operating costs (such as handling and waste treatment), and by making an efficient use of resources, process efficiency—and therefore economics—can be improved [2].

A key area of green route investigation is in the application of biocatalysts. Biocatalysts are defined as biological substances which can act as catalysts to chemical reactions. Their description is not confined solely to enzymes, and can include cellular organelles (a sub-unit with a cell), microbes, and plant and animal cells [3, 4]. Due to their ubiquitous nature, enzymes have largely been focused on in studies, and resultantly, the terms 'enzyme' and 'biocatalyst' are commonly used synonymously

A simplified explanation of how enzymes work is the 'lock and key' hypothesis. From this, only a specific substrate can fit into active sites on the enzyme to form an enzyme-substrate complex (Fig. 4.1). This complex therefore makes the substrate in a better position for reacting, resulting in a lower activation energy, and therefore leading to greater reaction rates. Enzymes are thus advantageous due to their reactant (substrate) specific catalysis, reduction in by-product formation and their inherent ability to catalyse reactions under mild conditions. This can therefore reduce operating costs and safety implications; they also improve the green credentials of processes. In some cases, enzymes can even make otherwise unattainable reactions

S. V. Patwardhan (✉) · C. Forsyth
Department of Chemical and Process Engineering, University of Strathclyde,
75 Montrose Street, G1 1XJ Glasgow, U.K.
e-mail: Siddharth.Patwardhan@strath.ac.uk

P. M. Zelisko (ed.), *Bio-Inspired Silicon-Based Materials,* Advances in Silicon Science 5,
DOI 10.1007/978-94-017-9439-8_4

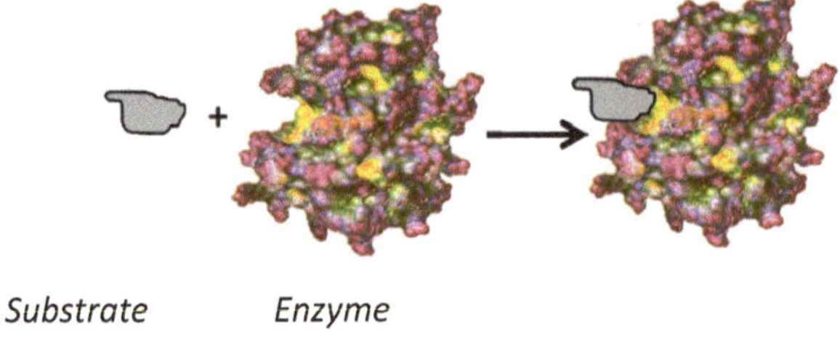

Fig. 4.1 Enzyme activity viewed as a simple 'lock and key' system

feasible [2]. It is therefore of great interest to make use of enzymes efficiently on an industrial scale. Indeed, presently, enzymes have found applications in biocatalysis, textile manufacturing, food processing, syntheses of pharmaceuticals, cosmetics, pollution control and anti-fouling coatings [5, 6]. This diversity of their current applications strongly supports the importance of enzymes.

Enzymes do, however, have several inherent limitations that restrict full utilisation of their potential on a commercial scale. One such problem is their soluble nature, which hinders the ease of enzyme recovery and reusability potential, and can also lead to instability. Resultantly, expensive and energy intensive separations are often necessary, and product contamination can be problematic. In addition, enzymes are often delicate and their molecular structure is prone to destruction under processing conditions [2, 7]. As a consequence, the 'lock and key' system can prove problematic due to the deformation of the shape of the enzyme—the substrate will no longer be able to attach to the active sites, therefore greatly reducing the effectiveness of the system. High temperatures, extreme pH conditions and the presence of other components such as organic solvents can all result in enzyme deformation (as well as causing detrimental effects through other means such as complicating chemistry). This therefore inhibits enzyme performance, and may even result in complete denaturation [3, 4, 8].

4.2 Enzyme Immobilisation Overview

To overcome the problems associated with enzymes, considerable research has been carried out on enzyme immobilisation techniques, whereby the enzymes are effectively supported/confined for the duration of the reaction. This allows their catalytic properties to be preserved and easier handling, while assisting with repeated and continuous use, therefore reducing operating costs [2–4]. Continuous operation is also preferable for improved throughput and efficiency [1, 2]. Effective immobilisation has also been shown to result in greater enzyme stability over more intense process conditions, such as more extreme temperatures and pH values, and non-aqueous conditions, thereby expanding their potential applications. Reactor design and control is also simpler for immobilised systems since the enzyme can

Table 4.1 Criteria for robust immobilised enzymes, adapted from ref. [9] with permission from Elsevier

Parameter	Requirement	Benefits
Non-catalytic	Suitable particle size	Ease of separation
	Easy fabrication	Flexibility in reactor design
	High mechanical and chemical stability	No decomposition, loss of support
Catalytic	High specific activity (U/g)	High productivity and space-time yield
	High selectivity	Less by-products, easy downstream processing
	Substrate specificity (broad)	Applicable to a wide substrate chemistry
	Thermal, chemical, conformational and operation stability	Low cost operation
Immobilised enzyme	Recyclability	Economical processing
	Reproducibility	Quality assurance
	Facile design	Short times for process development
Other	Low volume	Low handling and capital costs
	Easy disposal	Less pollution
	Safe for use	Safe

usually be utilised in fixed-bed type reactors or easily removed by simple filtration protocols [7].

In order to design and develop immobilised biocatalysts, the biocatalyst needs to perform "non-catalytic" and "catalytic" functions effectively and several criteria have been suggested for robust immobilised enzymes (Table 4.1) [9]. The catalytic functions include efficient catalysis, productivity, substrate/product selectivity and yield, while non-catalytic functions comprise of support's stability (mechanical, chemical and thermal), porosity and reusability. A range of methods have been outlined in the literature for improving the catalytic and non-catalytic functions of immobilised enzymes such as specific activity, stability and selectivity [3, 4, 9, 10] and are not detailed herein.

As knowledge about immobilised systems increases, so too do the applications. Current applications include the production of useful substances (particularly in the pharmaceuticals and food and drinks industries) and waste treatment (water in particular). In these areas, product purity is important, hence using immobilised enzymes can result in purer products since the product will be effectively enzyme free [8]. Furthermore, immobilisation of enzymes offers their use in the fabrication of analytical devices, as adsorbents in purifying proteins and enzymes, as a platform for performing solid phase protein chemistry and for controlled release of protein drugs [10].

4.2.1 *Outline of Enzyme Immobilisation Techniques*

There are many possible techniques available for enzyme immobilisation, however, finding new and improved techniques is an area of major research [7]. Enzyme immobilisation techniques can be classified into four categories: covalent binding, adsorption, cross-linking and entrapment/encapsulation. Although only briefly listed here, these protocols are discussed in detail in Sect. 4.3. Covalent attachment and adsorption are of a similar theme since the enzyme becomes attached through some form of interaction with a supporting material. This is possible because enzymes contain chemically active, ionic and/or hydrophobic groups. These functionalities are therefore able to interact with reactive sites on the support. These attachment methods are most common, however, more recent approaches are now beginning to take precedence [7, 11]. See Table 4.2 for schematic illustrations of each immobilisation technique.

Cross-linked enzyme aggregates, CLEAs, are a very different approach altogether since they do not necessarily involve a support, and instead make use of a bi-functional cross-linking agent which reacts with the enzyme to create aggregates. In entrapment and encapsulation, the enzyme is confined and trapped within a support, rather than attached through binding (Table 4.2) [7, 11].

4.2.2 *Supports*

Care must be taken when choosing a suitable supporting material, such as resins, glass, cellulose and silica. Materials used depend on the nature of the enzyme, type of immobilisation and reactions involved. The performance of a biocatalyst depends on both the enzyme itself, as well as the carrier properties as illustrated in Fig. 4.2.

Supporting materials should ideally have the following properties: [2, 3, 7, 12]

- A large surface area with good geometric congruence with the enzyme;
- A structure that does not detrimentally affect enzyme performance (i.e. minimal conformational hindrance);
- Sufficient chemical properties for immobilisation, such as adequate reactive groups;
- Adequate physical, chemical and biological stability for the required process conditions;
- Suitable strength and malleability;
- A low cost and be readily available.

Many of the factors mentioned come as a trade-off against each other, and therefore optimisation techniques to find optimal immobilisation arrangements are often

Table 4.2 Summary and schematic illustration of enzyme immobilisation techniques

Approach	Illustration	Description
Covalent Binding		Enzymes immobilised onto support through covalent bonds
Adsorption: Ionic Binding		Enzyme ionically bonded onto support through electrostatic interactions (dotted line)
Physical Adsorption		Enzyme physically adsorbed onto support through interactions such as van der Waals forces, hydrogen bonding and hydrophobic interactions
Cross-Linking/ Aggregation		Additional reagents cause inter-molecular cross-linking. If significant enough, enzymes become insoluble
Entrapment/ Encapsulation		Enzyme becomes entrapped in a surrounding lattice/ matrix structure or becomes 'encapsulated' in small capsule like structures

necessary [2]. In industry, pre-manufactured resin supports are commonly encountered due to their effectiveness, but their cost is a major hindrance. Manufactured ranges include Sepabeads, Amberlite, Lewatit and Eupergit [13]. They largely come in bead form due to their large surface area. Furthermore, a range of inorganic supports are commonly utilised and these include nanoparticles, gels and beads prepared from ceramics (silica, titania, calcium-compounds) and metals [14, 15].

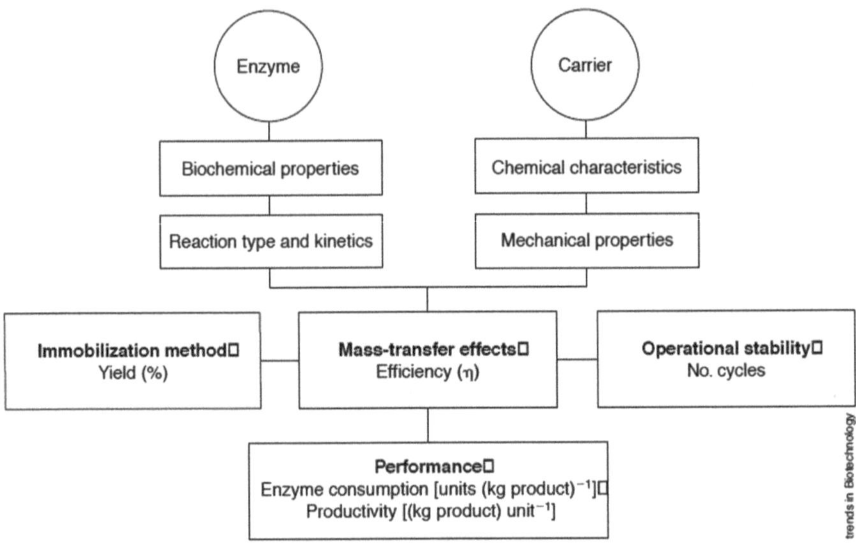

Fig. 4.2 Schematic representation of enzyme and carrier properties that influence the performance of immobilised enzymes. Figure reproduced from ref. [12] with permission from Elsevier

4.3 Immobilisation Techniques

This section considers each immobilisation technique in further details and compares with other existing methods. Although each approach has been adopted for a range of enzymes, for a clearer illustration and comparison, we have chosen lipase as an example enzyme. It is therefore important to outline key information of lipases such as properties and performance in biocatalysis prior to discussing their immobilisation. Before we discuss properties of lipase, we briefly present typical protocols used for measuring and comparing enzymatic activity.

4.3.1 Evaluation and Comparison of Biocatalyst Performance

There are many possible ways to evaluate and compare free and immobilised enzyme performance in terms of activity and stability, and this therefore makes direct comparisons between systems difficult at times. Therefore, it is often necessary to make qualitative, rather than quantitative comparisons.

To evaluate enzyme activity, a common approach is to measure the amount of product formed (or rate) over a fixed time duration and compare this between the free and immobilised systems. Another approach is to compare the enzyme kinetic parameters such as the maximum reaction rate (V_{max}), Michalis-Menton constant

(K_m) and catalytic rate constant (k_{cat}) for free and immobilised systems. The advantage of this approach is that the maximum rate is determined by performing analyses over a range of substrate concentrations, and the activity is therefore evaluated from a broader analysis, rather than a single-point condition. The majority of approaches, however, tend to favour the former, and only a small number of papers reviewed reported kinetic parameters [16–20]. Another difficulty in comparing enzyme activity is the use of normalised activity and a lack of absolute values of rates of product formation being reported.

Enzyme stability can be estimated in various ways. One way is to measure the enzyme performance/initial rates under varying conditions of temperature or pH for example [16, 21–24]. An alternative approach is to compare the half-life: the time it takes for enzyme activity to reduce to half of its original activity. The half-life is determined by monitoring the enzyme's activity at various durations, at a given condition, such as a temperature, which will result in a decrease in enzyme activity. Thermal enzyme denaturation typically follows an exponential decay, therefore allowing the half-life to be deduced, and big half-lives are indicative of better thermal stability. This approach allows a numerical increase in stability (the stabilisation factor) to be estimated [7, 25–27].

4.3.2 Lipase

Lipases are a technologically important family of enzymes within the esterase family. They are primarily associated with catalysing the hydrolysis of triglycerides (fats) into fatty acids and glycerol utilising a histidine-aspartate-serine catalytic triad. They are found in animals and plants, as well as micro-organisms such as bacteria and yeast [28]. Due to the ability to produce them on a large scale through micro-organism proliferation, they have been widely implemented on an industrial level in biotechnological applications such as the food processing industries (particularly dairy), pharmaceuticals industries (primarily to assist with producing enantiopureproducts) and chemicals industries (with significant use seen in producing synthetic polymers and detergents) [3–6, 11, 29, 30]. More recently, lipases have been implemented in the renewable energy industry to aid the conversion of oils to fuels, particularly in biodiesel production [31, 32]. This area is likely to see considerable growth in coming years.

The use of lipases on an industrial scale is, however, often limited due to reasons such as instability at higher temperatures and pH extremes. These drawbacks are particularly highlighted in the foods and pharmaceutical industries, where the production of pure, non-contaminated (enzyme-free) products is a necessity. The use of high temperatures is also often desirable in these industries, for reasons including improved reaction rates, improved mass transfer and higher levels of solubility [33, 34].

Due to the popularity of lipases in biotranformations, there are number of immobilisation methods which have been tested. Perhaps the most common form of

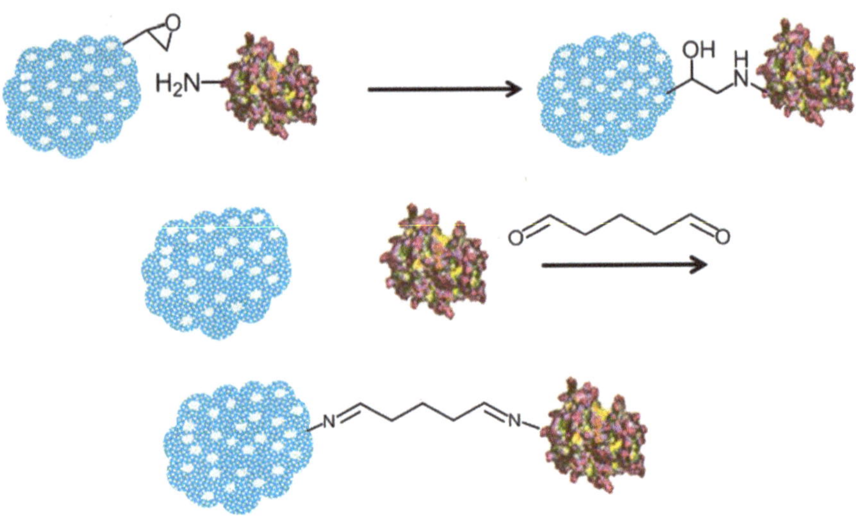

Fig. 4.3 Two examples of a covalent attachment through epoxy (top) or glutaraldehyde (bottom) chemistry [41]

immobilised lipase is the commercially available Novozym-435, which is produced by Novozymes. It consists of lipase *Candida antarctica B* (CALB) immobilised onto a macroporous acrylic resin called '*Lewatit VPOC 1600*' [35]. It is believed that lipase in Novozym-435 is physically adsorbed onto the resin through hydrophobic interactions, which is consistent with studies that suggest that Novozym-435 displays high levels of enzyme activity, but has poor stability and leaching problems [35–38].

4.3.3 Covalent Binding

With covalent binding, the reactive groups on the enzyme react with active sites on the support to form covalent attachments, therefore binding the enzyme and support together. The amine group on the enzyme is largely targeted in the reaction, however, thiol, phenolic and hydroxyl groups are often involved too. The supports are activated to render epoxy or glyoxyl reactive groups for enzyme attachment (Fig. 4.3) [7, 39, 40].

Covalent binding is perhaps the most common method of immobilisation, and as a result, there are many possible supports whose effectiveness have been demonstrated, including glass, resins, silica, zeolites, cellulose, chitosan, dextran, agarose and polyamides. Supports are often used in bead form due to their large surface area [42].

The chosen support must be activated to ensure that it has a suitable number of active sites to react with the enzyme. The most common method of activation is through the use of cyanogen bromide which creates highly reactive groups on the support, which react well with the amine groups on the enzyme. This therefore leads to the enzyme becoming covalently bound to the support [7, 43, 44]. Glutaraldehyde and aminopropyltriethoxysilane can also be used [45]. There are pre-activated supports commercially available, such as Sepabeads, which are composed of epoxy resins. These are frequently encountered in industry, and can be used to support a wide range on enzymes, including lipase, however, they are expensive [7]. Furthermore, naturally occurring supports such as the polysaccharides cellulose, chitosan, agarose and dextran are gaining popularity due to their low cost and availability [42, 46].

Compared to other methods, covalent binding has the benefit of strong covalent bonds (200–1000 kJ/mol) [47] between the support and the enzyme. This therefore results in minimal levels of leaks/leaching of enzyme from the support; this is beneficial since enzyme losses are minimal and product purity is maintained [48]. These strong bonds also regularly result in increased stability of the enzyme over a wider range of temperatures (and often pH), since the strong bonds help resist enzyme deformation. Covalent binding also results in high levels of contact between the enzyme and substrate since the enzyme is firmly localised on the surface of the support [41].

Covalent binding does, however, have its drawbacks. The strong bonds may result in conformational deformation of the enzyme, as well as hindering enzyme movement. Attachment of the enzyme to the support can also block active sites. These factors can result in reduced enzyme activity. The chemicals used for activation also mean that the route is usually not considered green [2]. Furthermore, the immobilisation process can be often time-consuming as illustrated with an example of mesoporous silica as a support in Fig. 4.4.

Glyoxyl agarose, as an example of covalent binding, has been reported to be an effective support for immobilising lipase. Agarose is a suitable support due to its high surface area and adequate porosity, and can be activated with epoxide reagents to create glyoxyl groups on the surface [26]. Reports suggest that 38–86 % of the enzyme's activity can be retained during immobilisation, and significant improvements in enzyme stability compared to the free enzyme have been seen, in terms of pH and temperature [7, 26, 50]. Stabilisation factors of up to 150 have been reported [7].

Commercial Eupergit resin support beads are also successful for lipase immobilisation, and they typically bind to the enzyme through epoxy, carboxyl or sulfhydryl groups [7, 51]. Retention of up to 80 % of enzyme activity has been reported [52]. This, however, depends on immobilization conditions, such as ionic strength and the presence of metal chelates.

Recently, green coconut fibres have been demonstrated as suitable support for covalent binding of lipase. For this, the fibres were first activated through silanisation using 3-glycidoxypropyltrimethoxysilane (GPTMS). Improved enzyme stability was seen (up to 360 times that of the free enzyme), and it also showed very good storage potential, with only small drops in activity [25].

The use of genetic engineering has also been demonstrated as means for improving immobilisation of lipase. Lipase has been genetically modified to display a free cysteine at a defined position, which was then attached to a glass surface [53].

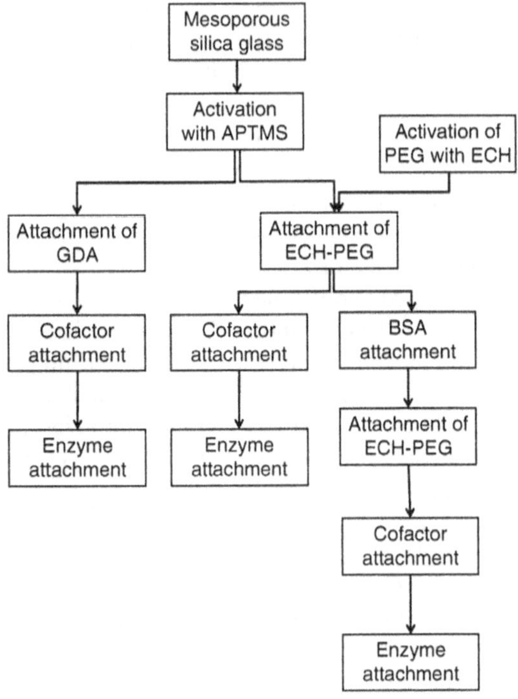

Fig. 4.4 A flow diagram of a typical multi-step process adopted for enzyme immobilisation on mesoporous silicas by covalent attachment. Reproduced with permission from Springer [49]

The engineered cysteine allowed immobilisation via the same site, thus keeping the same orientation of all immobilised enzyme molecules.

4.3.4 Adsorption

Adsorptive immobilisation of enzymes could involve ionic binding or physical adsorption through weaker interactions. For ionic binding, the enzyme is ionically bonded onto a support through electrostatic interactions between charged groups on the enzyme and the support. The charges on the enzyme depend on its conformation, as well as factors such as pH, ionic strength and temperature. Therefore, depending on these charges, a support with opposite charge can be appropriately selected [40]. In some cases, the support may need pre-treatment to optimise enzyme adsorption.

Many of the supports used for covalent binding can also be used for ionic binding, so long as they have suitable charges. For ionic binding, these supports are typically referred to as ion exchangers. Polystyrene supports have been shown to be effective for enzymes such as β-galactosidase, while polyethylenimine (PEI) has been proven effective for lipase immobilisation [8, 41, 54].

Ionic bonds are weaker than covalent (40–400 kJ/mol) [47], and therefore immobilisation results in less deformation to the enzyme. As a result, activity loss is typically less than that for covalent binding. In some cases, the process is also easily reversible under modified conditions, therefore allowing supports to be reused if necessary [2].

Due to the weaker bonds, however, leaching of enzyme from the support can be problematic, and this can lead to operational complexities (especially in highly ionic solutions), as well as reduced enzyme reuse potential. The weaker bonds also result in less stability than covalently bonded systems, and this therefore results in ionically bonded systems being less effective at more moderate processing conditions.

With physical adsorption, the enzyme is physically adsorbed onto the surface of the support through interactions such as van der Waals forces, hydrogen bonding and hydrophobic interactions. Which interactions are involved largely depends on the enzyme's conformation, as well as the support used. This method is viewed as a relatively simple route, since no chemical changes to the enzyme or support are necessary [40].

Supports used for covalent and ionic binding, which have strong adsorption capacities suitable for the enzyme in question, can be used. A key advantage of physical adsorption is that the supports generally require very little preparation. Resins, polyacrylates, polypropylene, polystyrene, metal oxides, activated carbon and silicas are common examples of suitable supports for a range of enzymes including lipase, amidase and amylase. Celite (diatomaceous earth) is popular for lipase in particular [41, 48, 55–57].

The physical interactions are the weakest of the attachments methods (4–40 kJ/mol) [47], and as a result, this method typically results in least enzyme deformation, thus high activities are maintained. As a consequence, a drawback of these weak interactions is that as with ionic binding, leaching can be problematic, especially since the interactions are easily reversed. This can therefore greatly restrict reuse potential of the enzyme. On the other hand, the supports can be easily regenerated by removing the spent enzyme. Again, a lack of stability is also a problem at more moderate processing conditions [39]. Due to its weaknesses, this method is often combined with cross-linking methods which are subsequently mentioned.

Lipase has been successfully immobilised via ionic binding to PEI coated agarose, with very high levels of activity being maintained (90 %). When immobilised under optimal conditions, the immobilised material retained its optimal performance, even when used in non-optimal conditions [41, 58].

Physical adsorption of lipase onto a celite support has been shown to result in very good product stability, with only negligible leaching. The preparation is also simple, with the enzyme simply being precipitated with celite powder. Additives such as sugar or albumin may be added to improve stability further [41, 59].

Polystyrene resins have also been demonstrated as suitable supports for lipase immobilisation via physical adsorption. Polystyrene resins are widely used to immobilise lipase, due to their simple preparation and high adsorption capacity. CALB also has a high affinity for the polystyrene, resulting in rapid immobilisation. Lipase immobilised in such a way has been shown to retain high levels of activity, and demonstrate improved stability [37, 42].

Lipase has been adsorbed onto highly hydrophobic octyl-agarose supports with success. Very high levels of lipase activity have been observed (with claims of 'hyperactivation' such that the immobilised lipase displays greater activity than free enzyme), and immobilisation occurred rapidly, making the method favourable. The strength of lipase adsorption was also strong enough to allow efficient use under mild conditions in a range of media (including organic solvents), and an improvement in thermal stability compared to free enzyme was observed [16].

The physical adsorption of lipase onto mesoporous silica nanoparticles has also been demonstrated. Lipase was physically adsorbed to SBA-15 silica spheres through hydrophobic interactions, by placing the spheres in a lipase solution. The silica nanoparticles used were lengthy to produce and entailed harsh conditions (temperatures of 550 °C). The activity maintained depended on pore size, with larger pores (24 nm) resulting in higher activity than smaller pores (5 nm) [17, 18, 60]; the activity of the lipase immobilized in larger pores was very similar to that of the free enzyme, whereas it was just 20–30 % in smaller pores. This was thought to be due to mass transfer limitations associated with smaller pore sizes [17]. Further studies suggest that for similar routes, the stability of the lipase can be improved by chemically pre-treating the enzyme with ethylene glycol bis (succinimidyl succinate) or glutaraldehyde; the half-life of glutaraldehyde treated lipase was 8 times that of non-treated lipase [27].

Lipase has also been adsorbed onto fumed silica supports. Fumed silica has a large surface area and high absorption capacity for proteins, therefore making it a suitable support for enzymes, and adsorption is thought to occur through electrostatic interactions. Adsorption was found to depend on factors such as pH, ionic strength and silica availability. The activity of the material compared well with the commercial Novozym-435, however, the thermal stability was poorer, and the material had less storage potential (in organics) [19].

4.3.5 Cross-Linking

With cross-linking, additional reagents are added to bring about inter-molecular cross-linking, and if significant enough, the enzyme becomes insoluble. First of all, a precipitating agent is added to precipitate the enzyme out of solution to create physical enzyme aggregates. These are, however, unstable and would readily dissolve back into solution. A cross-linking agent is therefore added to cross-link these physical enzyme aggregates, and it works by reacting with the reactive groups on the enzyme, therefore creating cross-linked enzyme aggregates (CLEA) [48]. Suitable chemicals to bring about physical aggregation include salts and organic solvents, such as acetone and ethanol [41]. Glutaraldehyde is the most common cross-linking agent, and is suitable for immobilising a range of enzymes including amylase, lipase, invertase, nitrilases and hydrolase [2, 61]. Alternative cross-linking agents include dimethyl suberimidate, toluene diisocyanate and hexamethylene diisocyanate; these are hazardous chemicals, and therefore means the process cannot be considered green. Additives such as bovine serum albumin and diamino compounds can also be added to enhance the procedure [10, 41, 62].

Cross-linking has the advantage of not requiring a support, and the aggregates formed also demonstrate high stability, making them suitable for use in more extreme processing conditions. They have also demonstrated good reuse potential [10]. However, cross-linking can result in conformational changes, therefore reducing enzyme performance. The nature of the aggregates can also result in reduced mass transfer, thus lessening the effectiveness of the system. The diversity and robustness of its applications are also limited due to the variable nature of enzymes, since some enzymes may resist cross-linking [2]. Although involving relatively simple procedures [48], the two-step method of precipitation and cross-linking may not be considered efficient. Cross-linking can be combined with methods involving supports, therefore allowing the advantages of supports to be implemented, and resulting in more effective systems [10].

Immobilisation through cross-linking of lipase has recently been carried out using glutaraldehyde as the cross-linking agent, and bovine serum albumin as an additive to improve stability. The immobilized lipase showed very good stability at high temperatures, and also in organic solvents. It also showed very good reuse potential, with only a small drop in activity after ten cycles [63].

Another study that used glutaraldehyde as the cross-linking agent found that the precipitating agent had a major impact on the lipase activity retained, as well as the yield of enzyme immobilised. For example, polyethylene glycol 200 resulted in the immobilized material having a higher specific activity than when acetone, butanone, propane, polyethylene glycol 600 and ammonium sulphate were used as precipitating agents, while ammonium sulphate and polyethylene glycol 600 resulted in the highest yields of immobilised lipase [20].

Lipase immobilized through CLEAs has also been shown to display better stability than Novozyme 435, particularly in organic solvents; this includes less leaching, therefore increasing reuse potential [64]. There have also been claims of lipase 'hyperactivation' for particular preparations-these made use of additives such as surfactants or crown ether, and ammonium sulphate, acetone or dimethoxyethane as precipitating agents [65].

4.3.6 Entrapment/Encapsulation

Although involving a support, entrapment and encapsulation methods do not require direct interactions between the enzyme and support, and instead make use of the support to confine and trap the enzyme. With entrapment, the enzyme becomes entrapped in a surrounding lattice/matrix structure, while with encapsulation, the enzyme becomes encapsulated in small capsule like structures.

Within these approaches of immobilisation, the sol-gel method is by far the most common approach and this involves immobilising the enzyme within a gel,

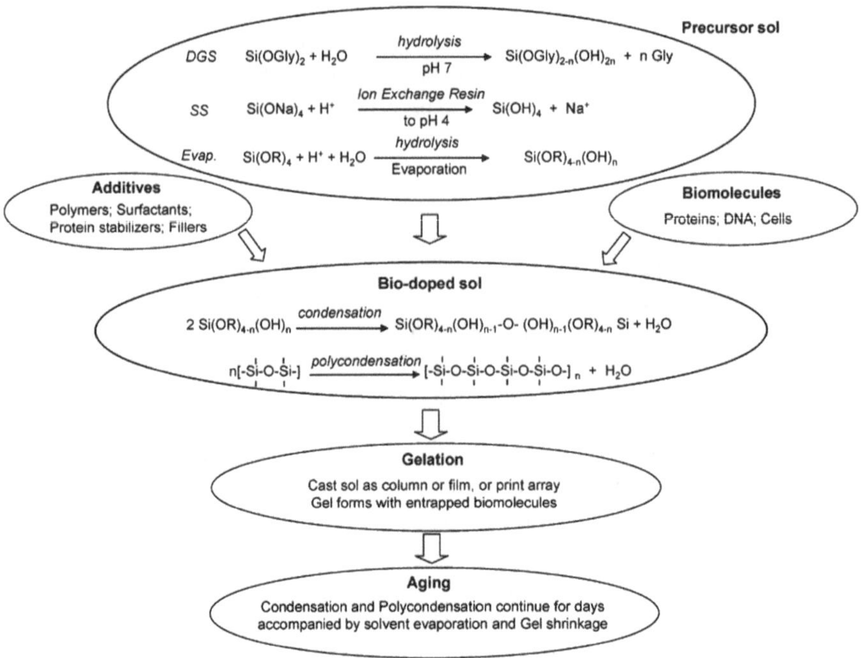

Fig. 4.5 Typical sol-gel process adopted for enzyme immobilisation. Image reproduced with permission [68]. *DGS* = diglycerylsilane; *SS* = sodium silicate; *Evap.* = alcohol removal by evaporation

typically silica [2, 10, 41, 66]. Alternative approaches of entrapment and encapsulation include hydrogels such as polyacrylamide gels [7, 67], immobilising the enzyme by using fibres, using semi-permeable membrane microcapsules or usinf polyelectrolyte micro-capsules [40].

The sol-gel route is, however, very common and this is largely due to its versatility, since it is compatible with a vast range of enzymes, including lipase, catalase and peroxidase [68]. We have recently reviewed elsewhere this method for the formation bio-hybrids [69] and a brief summary is provided in the following. It involves the hydrolysis and subsequent condensation and aggregation of alkoxysilanes, resulting in the creation of a gel or particulate-like silica network [2, 70]. The enzyme is added during this, resulting in its entrapment within the gel (Fig. 4.5). For entrapment/immobilisation methods in general, a major advantage is that modifications to the enzymes are minimal since the support simply confines the enzyme to restrict its movement, and as a result, high activities can be maintained. Since no direct attachment of the enzyme to the support occurs, active sites do not become blocked by the support, as is a problem for the attachment methods. Enzyme stability is also typically increased due to the surrounding shielding structure; this therefore increases applications at more extreme conditions.

Table 4.3 Examples of a wide variety of lipase immobilisation techniques

Method	Support/Cross-Linking Agent	Ref.
Covalent binding	Glyoxyl agarose	[7, 26, 50]
	Commercial Eupergit support beads	[7, 51, 52]
	Green coconut fibres	[25]
	Glass coated in polyethylene glycol	[53]
Ionic binding	PEI coated agarose	[41, 58]
Physical adsorption	Celite	[41, 59]
	Polystyrene resins	[37, 42]
	Octyl-agarose	[16]
	Silica mesoporous nanoparticles	[17, 18, 60]
	Fumed silica	[19]
Cross-linking	Glutaraldehyde	[20, 64]
Entrapment	Sol-gel Tetraethoxysilane (TEOS), tetramethoxysilane (TMOS), propyltrimethoxysilane (PTMS)	[21, 66 ,71– 73]
	Agarose and poly (N-isopropylacrylamide) hydrogels	[24]
	Siliceous mesocellular foam	[22]

Disadvantages of this system include a reliance on a porous network, which can result in reduced mass transfer, since it is often difficult for large substrate molecules to gain access to the enzyme due to complex, tortuous networks. This can greatly restrict the performance of the system [42].

The sol-gel route can be used to tune porosities of silica gels and thus allows mass transfer to be controlled. Sol-gel routes also occur in fairly benign conditions, typically at less than 100 °C [70]. However, sol-gel chemistry when used for enzyme immobilisation suffers from non-biocompatible precursors, production of alcohols and gel shrinkage—all causing enzyme deactivation.

In order to render sol–gel synthesis biocompatibility, several modifications to the process were successfully demonstrated. The formation of alcohol was avoided by one such modification on the precursor—the synthesis of biocompatible glycerated silanes or the use of water-soluble silicates.[68] The second improvement was realised when stabilising agents such as glycerol, carbohydrates or amino acids were added to the sol-gel process. A third modification was the introduction of multistep procedures wherein a pre-formed silica sol would be added to a buffered solution of enzyme in order to avoid any detrimental effects on the enzyme caused by pH drifts.

Despite these improvements, sol–gel methods still suffer from several issues including gelation times of several hours to weeks; the requirement of high silicate concentrations; tedious multistep protocols; and the requirement of custom-synthesised biocompatible precursors [69].

A recent approach involved impregnating a resin support with sol-gel precursor and enzyme (lipase). After gelation, the resultant immobilised enzyme displayed

Table 4.4 Key aspects to be considered for enzyme immobilisation, adapted from Reference [41]

Technique	Aspects to consider
Any	Interfering compounds in the formulation Enzyme stability during immobilisation conditions Stability of the carrier and enzyme under operating conditions Support-substrate interactions Availability and cost of the support/carrier
Adsorption/deposition	Presence of compatible regions (e.g. hydrophobic/hydrophilic) on enzyme
Ionic interactions	Enzyme pI, and pH and ionic strength of the immobilisation solvent Surface charged residues on the enzyme
Covalent binding/ cross-linking	Location of the bonding/linking residues Conditions of immobilization
Encapsulation	Size of 'cavity' and the enzyme Synthesis and immobilisation conditions

improved activity and stability compared with individual methods. Therefore, as with cross-linking, enhancements may be made by combining approaches [41].

Agarose and poly (*N*-isopropylacrylamide) hydrogels have also been successful at entrapping lipase. The entrapped lipase maintained good activity, and displayed very good reuse potential and improved stability compared to free enzyme, particularly in organics [24].

Lipase has also been entrapped in a siliceous mesocellular foam, which made use of a pressure-driven mechanism to bring about immobilisation, rather than conventional stirring. The immobilised material displayed high activity, little leaching and improved thermal stability compared to free enzyme. The pressure-driven mechanism also resulted in higher enzyme loadings and reduced leaching (and therefore improved reuse potential) than stirring [22].

In summary, a variety of immobilisation techniques are available and we have discussed each by using lipase as an example. These examples are presented in Table 4.3 for a quick overview. It is clear that for immobilisation, the choice of support/carrier used for enzyme immobilisation as well as the techniques employed depend on the enzyme, substrate, product and operating conditions and as such, one generic rule cannot be applied [41]. In addition, several considerations need to be taken into account when selecting immobilisation supports and techniques and they are listed in Table 4.4. As outlined, many of the current approaches to immobilisation suffer from several problems such as long preparation times, tedious multistep procedures, use of toxic/hazardous chemicals and loss of enzyme activity during immobilisation. Due to the materials and techniques involved, many could also not be classed as green. These drawbacks therefore restrict their industrial applications [7]. There does, however, appear to be much scope for improved enzyme immobilization techniques through the use of bioinspired silica. Due to its mild, rapid and controlled protocol, biologically inspired silica has recently been demonstrated as a green method for enzyme immobilisation via *in situ* entrapment of enzymes [69, 74, 75].

4.4 Bioinspired Silica for Enzyme Immobilisation

Biology produces nanostructured silica under environmentally-friendly green conditions. Extensive research has been undertaken in order to successfully find ways to transfer the bio-based methods into synthetic approaches to establish bioinspired silica technology. These methods, for inorganic materials synthesis, are green, rapid, controlled, mild and do not require non-aqueous solvents. Furthermore, such bioinspired synthesis allows material properties to be explicitly tailored.

Mild synthetic conditions of bioinspired silica formation have allowed the formation of nanostructured silicas containing peptides, proteins and enzymes [69]. A key advantage of using bioinspired method is that the materials properties (e.g. surface area, particle size, porosity, etc.) can be fine-tuned through the use of appropriate processing parameters, additives and silica precursors. This rapid, one-step, *in situ* method, when applied to enzyme immobilisation route, has great potential over existing routes due to its short preparation time, mild conditions and, as subsequently detailed, its flexibility [2, 69]. An extensive list of enzymes immobilised using bioinspired silica supports is tabulated in Table 4.5.

Initial studies concentrated on using the R5 peptide, which is a synthetic silica precipitating peptide, based on a naturally occurring silaffin protein found in the silica skeleton of the *Cylindrotheca fusiformis* diatom, as an additive. A range of enzymes including catalase, horseradish peroxidase and β-Galactosidase have been successfully immobilised using the R5 peptide, with all reporting very high levels of activity being maintained and good enzyme stability (Table 4.5) [45, 76]. Although, R5 peptide proved successful in immobilising a range of enzymes, the peptide is, however, expensive to obtain [2, 74].

From Table 4.5 it can be seen that several different additives, most of which are cheaper than R5 and readily available, have been used to produce silica capable of immobilizing a host of enzymes, all with varying success. Polyethyleneimine (PEI) has had varying success. Poor results were observed for lipase and carboxylesterase immobilisation, with significant loss of activity being observed compared to free enzyme [77, 78]. It did, however, appear efficient for horseradish peroxidase immobilisation [79]. Polyallylamine (PAH) has also had reasonable results. It has been used to successfully immobilise lipase and D-amino acid oxidase, with 51 and 96 % of the enzyme's activity being maintained respectively [80, 81]. Polyamidoamine dendrimers (PAMAM) and poly-L-lysine (PLL) have also proved successful in immobilising several enzymes. PAMAM has been used to immobilise glucose oxidase and horseradish peroxidase (with limited success) [79, 82], while PLL has been used to immobilise several enzymes including adenosine deaminase [83].

As is evident from Table 4.5, as well as for chemical processes, the immobilised enzymes produced using these methods have the potential to be used in other application, such as for use in biosensors, microfluidic devices and as anti-fouling coatings; many applications also involve integrating the immobilised enzymes onto surfaces [2, 45, 84]. This range of applications is likely to contribute to an increase in interest towards these bioinspired routes.

Table 4.5 A list of enzymes immobilised using bioinspired silica

Enzyme	Additive	Comment	Activity[a]	Ref.
Butyrylcholinesterase	R5	The immobilised enzyme appeared to retain virtually all of its activity. An improvement in stability was also observed	>90	[74]
Catalase; Horseradish peroxidase	R5	Both enzymes showed good stability, storage potential and activity similar to the free enzyme	90 90	[76]
β-Galactosidase	R5	Enzyme immobilised in silica nanospheres were attached to a silicon support, with very little loss of activity detected. Biocatalysts showed very good storage potential	~90	[2, 45]
Hydroxylaminobenzene mutase (HABM); Soybean peroxidase (SBP)	R5	The performance of immobilised enzymes was successfully tested in a microfluidic device. Greater levels of activity were maintained for SBP than HABM. The immobilised materials exhibited good reuse potential	45–65 65–85	[89]
Lipase; Glucose isomerase	PEI	Immobilised lipase retained very low levels of activity, while glucose isomerise was reasonably active. Poor lipase activity was thought to be due to PEI creating a strongly hydrophilic environment, making substrate access difficult. Lipase did, however, showed improved thermal stability and a wider pH optimum	10 40	[2, 77]
Horseradish peroxidase	PEI PAMAM	Immobilised enzymes were used for biosensor applications. PEI appeared most efficient for silica formation and the immobilised enzyme was effective for use in a biosensor	N/S	[79]
Carboxylesterase	Lysozyme R5 PEI	Both lysozyme and R5 proved successful; however, immobilisation was unsuccessful when PEI was used	N/S	[78]
Hexose oxidase	PEI	Enzyme was tested for potential use as enzyme based anti-fouling coatings. Immobilised enzyme showed improved stability, including in an artificial seawater environment, as well as good long term properties	6–13[b]	[85]
Luciferase	Cysteamine Ethanolamine	Higher levels of activity were maintained using cysteamine, rather than ethanolamine. It was therefore deduced that this was due to cysteamine having better catalytic properties for silica formation	87 17	[90]

Table 4.5 (continued)

Enzyme	Additive	Comment	Activity[a]	Ref.
Lipase (Pseudomonas cepacia)	PAH	Good levels of activity were maintained compared to free enzyme. Activity was enhanced further (up to 3 times) by using alkyl-alkoxysilanes. Improved stability and good storage potential was seen	51	[80]
Acetylcholinesterase; Glucose oxidase; β-Galactosidase	PEI	Maximum rates (and therefore activity) of immobilised system was less than free enzyme system. Immobilised enzyme also needed higher substrate concentrations than free enzyme	40[b]	[87]
Adenosine deaminase; Nucleoside phosphorylase; Xanthine oxidase	PLL	The immobilised enzymes were used to produce biosensors. The biosensors produced performed well, with high performances, suggesting immobilisation had little detrimental effect on the enzymes. The procedure was also adaptable for other enzymes	N/S	[83]
β-glucoronidase	Protamine	Activity of immobilised system was less than free enzyme system. Leaching was negligible, an improvement in stability was seen, and the material had excellent reuse and storage potential	20–30[b]	[86]
Glucose oxidase; Horseradish peroxidase	PAMAM	Good levels of glucose oxidase activity were retained during immobilisation in water (better than horseradish peroxidase). It showed very little leaching and good storage potential. Horseradish peroxidase performed better that glucose oxidase when immobilised in buffer	4–54 13–39	[82]
D-amino acid oxidase	PAH	High levels of enzyme activity were maintained upon immobilisation. An improvement in stability was seen, compared to free enzyme	96	[81]
Papain	PLL-*b*-PLG[c]	The immobilised enzyme retained good activity and demonstrated enhanced pH and thermal stability, as well as good reuse potential. With this approach, the papain was first entrapped in the block copolypeptide vesicles, and added to a silica precursor to initiate silica formation	~40[b]	[88]
Invertase	PEHA	First report presenting enzyme immobilisation using flow chemistry useful for continuous production. Enzyme was observed to be active upon immobilisations	N/S	[91]

N/S not stated

[a] Activity % cf. Free Enzyme

[b] Activity estimated from kinetic parameters

[c] Poly-L-lysine–*b*-poly-L-glycine copolymer

Comparing the effectiveness of different systems is, however, difficult since due to mass transfer being a major limitation, the success of the system depends very much on the size of the substrate used in testing the material [2]. As with the lipase examples, enzyme activity is reported in various ways, with only a few papers determining kinetic parameters [81, 82, 85–88], and the majority making use of a single-point estimation for a given condition. Different reporting methods are also used for depicting enzyme stability.

4.5 Conclusion

Enzymes undoubtedly play a major role in industry, and their popularity is likely to grow over coming years due to increasing demand for more environmentally friendly and efficient processes. Many of the current methods of enzyme immobilisation have significant drawbacks, such as greatly reducing enzyme activity, being costly and time-consuming to prepare, and not having 'green' credentials due to materials and conditions involved.

The bioinspired silica immobilisation routes outlined, however, aim to surmount these challenges, with simple procedures, mild conditions, short preparation times and the use of low cost additives. As a result, research and investment in this area of enzyme immobilisation is likely to expand in the future. Moving forward, one should note some of the issues with bioinspired silica for enzyme immobilisation. Some of these challenges include the so far lack of any demonstrable control on mass transport of substrate/product to and from the immobilised enzyme. This issue perhaps can be tackled through silica chemistry by systematically tuning porosities, particle sizes and morphologies of the support. Availability of complete characterisation of immobilised enzymes—support properties as well as enzyme kinetic parameters—has been poor in many reports, thereby hindering the process of mechanistic and molecular level understanding. In order for the potential of this technology to be commercially realised, investigations focussing on process development would be of immense importance.

4.6 Acknowledgments

The authors acknowledge the financial support from the Department of Chemical and Process Engineering, the Faculty of Engineering at University of Strathclyde, the Royal Society (grant number TG090299) and The Carnegie Trust Vacation Scholarship. SVP thanks Professor S. J. Clarson and his group for various helpful discussions.

References

1. Anastas PT, Zimmerman JB (2003) Design through the 12 principles of green engineering. Environ Sci Technol 37(5):94A–101A
2. Betancor L, Luckarift HR (2008) Bioinspired enzyme encapsulation for biocatalysis. Trends Biotechnol 26(10):566–572
3. Aehle W, Perham RN, Michal G, Caddow AJ et al (2008) Enzymes, 1. General. Ullmann's encyclopedia of industrial chemistry. Wiley-VCH Verlag GmbH & Co. KGaA. doi:10.1002/14356007.a09_341.pub3, Weinheim
4. Aehle W, Perham RN, Michal G, Jonke A et al (2003) Enzymes. Ullmann's encyclopedia of industrial chemistry. Wiley-VCH Verlag GmbH & Co. KGaA. doi:10.1002/14356007.a09_341.pub2, Weinheim
5. Kirk O, Borchert TV, Fuglsang CC (2002) Industrial enzyme applications. Curr Opin Biotechnol 13(4):345–351
6. Langrand G, Rondot N, Triantaphylides C, Baratti J (1990) Short chain flavor esters synthesis by microbial lipases. Biotechnol Lett 12(8):581–586
7. Mateo C, Palomo JM, Fernandez-Lorente G, Guisan JM et al (2007) Improvement of enzyme activity, stability and selectivity via immobilization techniques. Enzyme Microb Technol 40(6):1451–1463
8. Bailey JE, Ollis DF (1986) Biochemical engineering fundamentals. 2nd edn. McGraw-Hill, New York
9. Cao LQ, van Langen L, Sheldon RA (2003) Immobilised enzymes: carrier-bound or carrier-free? Curr Opin Biotech 14(4):387–394
10. Cao LQ (2005) Immobilised enzymes: science or art? Curr Opin Chem Biol 9(2):217–226
11. Balcao VM, Paiva AL, Malcata FX (1996) Bioreactors with immobilized lipases: State of the art. Enzyme Microb Technol 18(6):392–416
12. Tischer W, Kasche V (1999) Immobilized enzymes: crystals or carriers? Trends Biotechnol 17(8):326–335
13. Brady D, Jordaan J (2009) Advances in enzyme immobilisation. Biotechnol Lett 31(11):1639–1650
14. Weetall HH (1993) Preparation of Immobilized Proteins Covalently Coupled Through Silane Coupling Agents to Inorganic Supports. Appl Biochem Biotechnol 41(3):157–188
15. Weetall HH (1969) Trypsin and papain covalently coupled to porous glass: preparation and characterization. Science 166:615–617
16. Bastida A, Sabuquillo P, Armisen P, Fernandez-Lafuente R et al (1998) Single step purification, immobilization, and hyperactivation of lipases via interfacial adsorption on strongly hydrophobic supports. Biotechnol Bioeng 58(5):486–493
17. Jaladi H, Katiyar A, Thiel SW, Guliants VV et al (2009) Effect of pore diffusional resistance on biocatalytic activity of Burkholderia cepacia lipase immobilized on SBA-15 hosts. Chem Eng Sci 64(7):1474–1479
18. Zheng S (2005) Effect of Pore Curvature and Surface Chemistry of Model Silica Hosts on Biocatalytic Activity if Immobilised Lipase. (Electronic Thesis or Dissertation). Accessed https://etd.ohiolink.edu/. University of Cincinnati, Cincinnati.
19. Cruz JC, Pfromm PH, Rezac ME (2009) Immobilization of Candida antarctica Lipase B on fumed silica. Process Biochem 44(1):62–69
20. Devi B, Guo Z, Xu XB (2009) Characterization of cross-linked lipase aggregates. J Am Oil Chem Soc 86(7):637–642
21. Reetz MT, Zonta A, Simpelkamp J (1996) Efficient immobilization of lipases by entrapment in hydrophobic sol-gel materials. Biotechnol Bioeng 49(5):527–534
22. Han Y, Lee SS, Ying JY (2006) Pressure-driven enzyme entrapment in siliceous mesocellular foam. Chem Mater 18(3):643–649
23. He F, Zhuo RX, Liu LJ, Jin DB et al (2001) Immobilized lipase on porous silica beads: preparation and application for enzymatic ring-opening polymerization of cyclic phosphate. React Funct Polym 47(2):153–158

24. Bai S, Wu CZ, Gawlitza K, von Klitzing R et al (2010) Using hydrogel microparticles to transfer hydrophilic nanoparticles and enzymes to organic media via stepwise solvent exchange. Langmuir 26(15):12980–12987
25. Brigida AIS, Pinheiro ADT, Ferreira ALO, Pinto GAS et al (2007) Immobilization of *Candida antarctica* lipase B by covalent attachment to green coconut fiber. Appl Biochem Biotechnol 137:67–80
26. Rodrigues DS, Mendes AA, Adriano WS, Goncalves LRB et al (2008) Multipoint covalent immobilization of microbial lipase on chitosan and agarose activated by different methods. J Mol Catal B-Enzym 51(3–4):100–109
27. Forde J, Vakurov A, Gibson TD, Millner P et al (2010) Chemical modification and immobilisation of lipase B from Candida antarctica onto mesoporous silicates. J Mol Catal B-Enzym 66(1–2):203–209
28. Contesini FJ, Lopes DB, Macedo GA, Nascimento MdG et al (2010) *Aspergillus sp* lipase: Potential biocatalyst for industrial use. J Mol Catal B-Enzym 67(3–4):163–171
29. Houde A, Kademi A, Leblanc D (2004) Lipases and their industrial applications. Appl Biochem Biotechnol 118(1):155–170
30. Pandey A, Benjamin S, Soccol CR, Nigam P et al (1999) The realm of microbial lipases in biotechnology. Biotechnol Appl Biochem 29(2):119–131
31. Ban K, Kaieda M, Matsumoto T, Kondo A et al (2001) Whole cell biocatalyst for biodiesel fuel production utilizing *Rhizopus oryzae* cells immobilized within biomass support particles. Biochem Eng J 8(1):39–43
32. Schoerken U, Kempers P (2009) Lipid biotechnology: Industrially relevant production processes. Eur J Lipid Sci Technol 111(7):627–645
33. Matsuura S-i, Ishii R, Itoh T, Hamakawa S et al (2009) On-chip encapsulation of lipase using mesoporous silica: A new route to enzyme microreactors. Mater Lett 63(28):2445–2448
34. Hwang S, Lee KT, Park JW, Min BR et al (2004) Stability analysis of Bacillus stearothermophilus L1 lipase immobilized on surface-modified silica gels. BiochemEng J 17(2):85–90
35. Cabrera Z, Fernandez-Lorente G, Fernandez-Lafuente R, Palomo JM et al (2009) Novozym 435 displays very different selectivity compared to lipase from *Candida antarctica* B adsorbed on other hydrophobic supports. J Mol Catal B-Enzym 57(1–4):171–176
36. Chen B, Miller EM, Miller L, Maikner JJ et al (2007) Effects of macroporous resin size on *Candida antarctica* lipase B adsorption, fraction of active molecules, and catalytic activity for polyester synthesis. Langmuir 23(3):1381–1387
37. Chen B, Miller ME, Gross RA (2007) Effects of porous polystyrene resin parameters on *Candida antarctica* Lipase B adsorption, distribution, and polyester synthesis activity. Langmuir 23(11):6467–6474
38. Chen B, Hu J, Miller EM, Xie W et al (2008) *Candida antarctica* lipase B chemically immobilized on epoxy-activated micro- and nanobeads: Catalysts for polyester synthesis. Biomacromolecules 9(2):463–471
39. Bhatia SC (2008) Textbook of Biotechnology. Atlantic Publishing, New Delhi
40. Murty VR, Bhat J, Muniswaran PKA (2002) Hydrolysis of oils by using immobilized lipase enzyme: A review. Biotechnol Bioprocess Eng 7(2):57–66
41. Hanefeld U, Gardossi L, Magner E (2009) Understanding enzyme immobilisation. Chem Soc Rev 38(2):453–468
42. Tischer W, Wedekind F (1999) Immobilized enzymes: Methods and applications. Top Curr Chem 200:95–126
43. Wykesa JR, Dunnilla P, Lilly MD (1971) Immobilisation of α-amylase by attachment to soluble support materials. Biochim Biophys Acta (BBA)—Enzymology 250(3):522–529
44. Knight K (2000) Immobilization of lipase from Fusarium solani FS1. Braz J Microbiol 31(3):219–221
45. Betancor L, Luckarift HR, Seo JH, Brand O et al (2008) Three-dimensional immobilization of beta-galactosidase on a silicon surface. Biotechnol Bioeng 99(2):261–267
46. Chiou SH, Wu WT (2004) Immobilization of Candida rugosa lipase on chitosan with activation of the hydroxyl groups. Biomaterials 25(2):197–204

47. Averill BA, Eldredge P (2007) Chemistry: Principles, Patterns, and Applications 1st edition. Prentice Hall
48. Sheldon RA, Schoevaart R, van Langen LM (2006) Cross-linked enzyme aggregates. Methods in Biotechnol 22:31–45
49. Buthe A, Wu S, Wang P (2011) Nanoporous Silica glass for the immobilization of interactive enzyme systems. In: Minteer SD (ed) Enzyme stabilization and immobilization: Methods and protocols. Springer Protocols-Humana Press, New York
50. Otero C, Ballesteros A, Guisan JM (1988) Immobilization stabilization of lipase from CANDIDA-RUGOSA. Appl Biochem Biotechnol 19(2):163–175
51. Katchalski-Katzir E, Kraemer DM (2000) Eupergit ® C, a carrier for immobilization of enzymes of industrial potential. J Mol Catal B-Enzymatic 10(1–3):157–176
52. Barbosa O, Ortiz C, Torres R, Fernandez-Lafuente R (2011) Effect of the immobilization protocol on the properties of lipase B from *Candida antarctica* in organic media: Enantiospecifc production of atenolol acetate. J Mol Catal B-Enzym 71(3–4):124–132
53. Blank K, Morfill J, Gaub HE (2006) Site-specific immobilization of genetically engineered variants of *Candida antarctica* lipase B. ChemBioChem 7(9):1349–1351
54. Serra I, Serra CD, Rocchietti S, Ubiali D et al (2011) Stabilization of thymidine phosphorylase from *Escherichia coli* by immobilization and post immobilization techniques. Enzyme Microb Technol 49(1):52–58
55. Galarneau A, Mureseanu M, Atger S, Renard G et al (2006) Immobilization of lipase on silicas. Relevance of textural and interfacial properties on activity and selectivity. New J Chem 30(4):562–571
56. Daoud FB-O, Kaddour S, Sadoun T (2010) Adsorption of cellulase *Aspergillus niger* on a commercial activated carbon: Kinetics and equilibrium studies. Colloids Surf B-Biointerfaces 75(1):93–99
57. Reshmi R, Sanjay G, Sugunan S (2006) Enhanced activity and stability of alpha-amylase immobilized on alumina. Catal Commun 7(7):460–465
58. Torres R, Ortiz C, Pessela BCC, Palomo JM et al (2006) Improvement of the enantio selectivity of lipase (fraction B) from Candida antarctica via adsorpiton on polyethylenimine-agarose under different experimental conditions. Enzyme Microbial Techno 39(2):167–171
59. Veum L, Kanerva LT, Halling PJ, Maschmeyer T et al (2005) Optimisation of the enantioselective synthesis of cyanohydrin esters. Adv Synth Catal 347(7–8):1015–1021
60. Serra E, Diez E, Diaz I, Blanco RM (2010) A comparative study of periodic mesoporous organosilica and different hydrophobic mesoporous silicas for lipase immobilization. Microporous Mesoporous Mater 132(3):487–493
61. Kennedy JF, Kalogerakis B, Cabral JMS (1984) Surface immobilization and entrapping of enzymes on glutaraldehyde crosslinked gelatin particles. Enzyme Microb Technol 6(3):127–131
62. Moehlenbrock MJ, Minteer SD (2011) Introduction to the field of enzyme immobilization and stabilization. In: Minteer SD (ed) Enzyme stabilization and immobilization: Methods and protocols. Springer Protocols-Humana Press, New York
63. Gupta MN, Raghava S (2011) Enzyme stabilization via cross-linked enzyme aggregates. Methods mol biol 679:133–145
64. Sheldon RA (2007) Cross-linked enzyme aggregates (CLEA ® s): stable and recyclable biocatalysts. Biochem Soc Trans 35:1583–1587
65. Lopez-Serrano P, Cao L, van Rantwijk F, Sheldon RA (2002) Cross-linked enzyme aggregates with enhanced activity: Application to lipases. Biotechnol Lett 24(16):1379–1383
66. Tomin A, Weiser D, Hellner G, Bata Z et al (2011) Fine-tuning the second generation solgel lipase immobilization with ternary alkoxysilane precursor systems. Process Biochem 46(1):52–58
67. Gonzalez-Saiz JM, Pizarro C (2001) Polyacrylamide gels as support for enzyme immobilization by entrapment. Effect of polyelectrolyte carrier, pH and temperature on enzyme action and kinetics parameters. Eur Polym J 37(3):435–444

68. Brennan JD (2007) Biofriendly sol-gel processing for the entrapment of soluble and membrane-bound proteins: Toward novel solid-phase assays for high-throughput screening. Acc Chem Res 40(9):827–835
69. Patwardhan SV (2011) Biomimetic and bioinspired silica: recent developments and applications. Chem Commun 47(27):7567–7582
70. Brinker J, Scherer G (1990) Sol-gel science: The physics and chemistry of sol-gel processing. Academic Press, Boston
71. Reetz MT (1997) Entrapment of biocatalysts in hydrophobic sol-gel materials for use in organic chemistry. Adv Mater 9(12):943–954
72. Reetz MT (2006) Practical protocols for lipase immobilization via sol-gel techniques. Method Biotechnol 22:65–76
73. Reetz MT, Tielmann P, Wiesenhofer W, Konen W et al (2003) Second generation sol-gel encapsulated lipases: Robust heterogeneous biocatalysts. Adv Synth Catal 345(6–7):717–728
74. Luckarift HR, Spain JC, Naik RR, Stone MO (2004) Enzyme immobilization in a biomimetic silica support. Nat Biotechnol 22(2):211–213
75. Patwardhan SV, Clarson SJ, Perry CC (2005) On the role(s) of additives in bioinspired silicification. Chem Commun 9:1113–1121
76. Naik RR, Tomczak MM, Luckarift HR, Spain JC et al (2004) Entrapment of enzymes and nanoparticles using biomimetically synthesized silica. Chem Commun 15:1684–1685
77. McAuliffe JC, Smith WC, Bond R, Zimmerman J et al (2005) Template-driven enzyme immobilization: Development of a rapid and practical process inspired by diatoms. Abstracts of Papers of the American Chemical Society 229:U1159–U1159.
78. Edwards JS, Kumbhar A, Roberts A, Hemmert AC et al (2011) Immobilization of active human carboxylesterase 1 in biomimetic silica nanoparticles. Biotechnol Progr 27(3):863–869
79. Zamora P, Narvaez A, Dominguez E (2009) Enzyme-modified nanoparticles using biomimetically synthesized silica. Bioelectrochemistry 76(1–2):100–106
80. Chen G-C, Kuan IC, Hong J-R, Tsai B-H et al (2011) Activity enhancement and stabilization of lipase from Pseudomonas cepacia in polyallylamine-mediated biomimetic silica. Biotechnol Lett 33(3):525–529
81. Kuan IC, Wu J-C, Lee S-L, Tsai C-W et al (2010) Stabilization of D-amino acid oxidase from Rhodosporidium toruloides by encapsulation in polyallylamine-mediated biomimetic silica. Biochem Eng J 49(3):408–413
82. Miller SA, Hong ED, Wright D (2006) Rapid and efficient enzyme encapsulation in a dendrimer silica nanocomposite. Macromol Biosci 6(10):839–845
83. Tian F, Wu W, Broderick M, Vamvakaki V et al (2010) Novel microbiosensors prepared utilizing biomimetic silicification method. Biosens Bioelectron 25(11):2408–2413
84. Johnson GR, Luckarift HR (2011) Enzyme stabilization via bio-templated silicification reactions. Methods mol biol 679:85–97
85. Kristensen JB, Meyer RL, Poulsen CH, Kragh KM et al (2010) Biomimetic silica encapsulation of enzymes for replacement of biocides in antifouling coatings. Green Chem 12(3):387–394
86. Li L, Jiang Z, Wu H, Feng Y et al (2009) Protamine-induced biosilica as efficient enzyme immobilization carrier with high loading and improved stability. Mater Sci Eng CMater Biol Appl 29(6):2029–2035
87. Neville F, Broderick MJF, Gibson T, Millner PA (2011) Fabrication and activity of silicate nanoparticles and nanosilicate-entrapped enzymes using polyethyleneimine as a biomimetic polymer. Langmuir 27(1):279–285
88. Lai JK, Chuang TH, Jan JS, Wang SSS (2010) Efficient and stable enzyme immobilization in a block copolypeptide vesicle-templated biomimetic silica support. Colloids Surf B-Biointerfaces 80(1):51–58
89. Luckarift HR, Ku BS, Dordick JS, Spain JC (2007) Silica-immobilized enzymes for multistep synthesis in microfluidic devices. Biotechnol Bioeng 98(3):701–705
90. Roth KM, Zhou Y, Yang WJ, Morse DE (2005) Bifunctional small molecules are biomimetic catalysts for silica synthesis at neutral pH. J Am Chem Soc 127(1):325–330
91. Patwardhan SV, Perry CC (2010) Synthesis of enzyme and quantum dot in silica by combining continuous flow and bioninspired routes. Silicon 2:33–39

Chapter 5
On The Immobilization of *Candida antarctica* Lipase B onto Surface Modified Porous Silica Gel Particles

Stephen J. Clarson, Richard A. Gross, Siddharth V. Patwardhan and Yadagiri Poojari

5.1 Introduction

A variety of lipases have proven to be versatile and efficient biocatalysts and have been utilized in transformations for esterification, transesterification, and ester hydrolysis reactions. Lipases have been widely employed in the food [1, 2], perfume [3], and detergent [4] industries. Lipases were also utilized as a catalyst in polymer synthesis [5], for such systems as polyesters [6–8] and polycarbonates [9, 10]. The vast potential of these biocatalysts for use in industrial applications has been increasingly recognized due to the high chemical selectivity, region selectivity, and stereo selectivity of lipases under mild reaction conditions.

In recent years, the search for efficient methods and support materials for the immobilization of lipases has emerged as an important field of interest. This is mainly due to the need to recover and reuse the enzyme for commercial benefits. A variety of techniques have been applied to immobilize lipases, including adsorption onto solid supports [11], covalent attachment [12–14], and entrapment within polymers/inorganic matrices [15, 16]. In general, adsorption techniques are easy to perform, but the binding of the enzyme is often weak and such biocatalysts lack the degree of stabilization that it is possible to attain by entrapment or covalent attachment. On the other hand, the enzyme entrapment leads to permanent loss of active sites

S. J. Clarson (✉) · Y. Poojari
Department of Chemical and Materials Engineering and the Polymer Research Centre, The University of Cincinnati, Cincinnati, OH 45221, USA
e-mail: stephen.clarson@uc.edu

R. A. Gross
Department of Chemistry and Biology, Rensselaer Polytechnic Institute, Troy, NY 12180, USA

S. V. Patwardhan
Department of Chemical and Process Engineering, University of Strathclyde, Glasgow, G1 1XJ, UK

P. M. Zelisko (ed.), *Bio-Inspired Silicon-Based Materials,* Advances in Silicon Science 5,
DOI 10.1007/978-94-017-9439-8_5

that are buried deep inside the matrix material. However, silica-based carriers have widely been used for enzyme immobilization by adsorption, entrapment or covalent attachment due to their inertness and stability under a wide variety of reaction conditions [17].

Several authors [11, 18] have reported data on the deposition of (3-aminopropyl) triethoxysilane (3-APS or γ-APS) on the surface of silica particles under various conditions. These include solvent, heat, and time, and then exposed to different curing environments, including air, heat, and ethanol. The authors found that 3-APS monolayers were formed under very mild reaction and curing conditions (reaction in dry toluene for 15 min at room temperature, curing in air, or 15 min in 200 °C oven), whereas thick layers were formed when reaction and curing time was increased. The 3-APS initially adsorbed onto the surface, and curing was found to be necessary to complete the covalent bond formation between the 3-APS and the hydroxyl groups present on the silica surface. Deposition of 3-APS from aqueous solution gave thin layers, due to water molecules being electrostatically bound to the silica. About 0.4 % APS in solution was the maximum concentration which still allowed monolayer coverage without physisorption. The monolayer films were deposited from dry organic solvent, cured under mild conditions, and incubated in water before use. However, thick films were deposited from an aqueous medium, cured in an oven or exposed to an open atmosphere for several days before use.

He et al. [19] reported that porcine pancreas lipase (PPL) immobilized on porous silica beads successfully catalyzed the ring-opening polymerization of ethylene isobutyl phosphate (EIBP). The porous silica beads were treated with methanesulfonic acid and then silanization using 3-APS. The lipase was covalently immobilized onto these functional porous silica beads by crosslinking with glutaraldehyde. It was found that the recovered immobilized lipase was more active for the polymerization of EIBP than for the first use. The number average molecular weight, M_n was found to be significantly increased.

In another study, Hwang et al. [20] reported that surface-modified silica gels were used as a carrier material for enzyme immobilization. Hydrophilic and hydrophobic silica gels were made by polyethyleneimine coating and silanization, respectively. Covalently bound lipase was found to be more stable than the lipase immobilized by physical adsorption. The surface chemistry of carriers was found to play a significant role on the stability of immobilized lipase, presumably through molecular interactions with lipase leading to structural effects in the enzyme.

Dragoi and Dumitriu [21] have reported that lipase B from *Candida antarctica* (CALB) was physically and chemically immobilized on mesoporous silica through functionalization and hydrophobization with 3-APS, chlorotrimethylsilane (CTMS), and propyltrimethoxysilane (PTMS). Physically immobilized CALB on the hydrophobized supports exhibited higher activities in comparison to covalently immobilized supports. However, the stability of the enzymatic preparations was improved by covalent immobilization and the biocatalyst exhibited significant activity after the third cycle of the alcoholysis of ethyl acetate with 1-hexanol and 1-butanol, respectively. Recently, an alternative approach to enzyme immobilization has been introduced, which relies on the implementation of mechanisms involved in biosilica formation [22]. This bioinspired method provides a green and one step process for lipase immobilization [23].

Among all of the lipases that have been reported in the literature for organic synthesis, CALB has proved to be the most efficient and highly stable in organic media during esterification and transesterification reactions [24, 25]. CALB is a globular protein with approximate molecular dimensions of 30 Å × 40 Å × 50 Å, and relative mass of 33 kDa [26]. CALB has a Ser–His–Asp catalytic triad in its active site when in an 'open' conformation with a restricted entrance. The active site is responsible for the substrate specificity and high degree of stereo-specificity of CALB, which is an α/β type protein that has many features in common with other lipase structures and related enzymes. However, the primary sequence has no significant homology to any other known lipase and it deviates from the sequence around the active site serine that is found in other lipases. CALB was physically immobilized within a macroporous resin (Lewatit VP OC 1600) poly(methyl methacrylate-*co*-divinylbenzene) and sold under the brand name Novozym-435®. The polymeric resin has a reported average particle size of 315–1000 μm, a surface area of 130 m^2g^{-1}, and a pore diameter of ~ 150 Å, respectively [15]. However, Novozym-435 has been reported to exhibit poor mechanical stability and the enzyme was found to be leaching out during the reactions carried out in organic media [24, 25].

In this chapter we report the immobilization of CALB using two approaches. Firstly, CALB was immobilised onto silica gel particles that had been surface modified porous by 3-APS, followed by chemically crosslinking with glutaraldehyde. The second approach adopted a bioinspired green method. The catalytic activity and the thermal stability of these particles were studied using an esterification model reaction between 1-octanol and lauric acid in isooctane. The results were then assessed and compared to those for free CALB and to the commercial Novozym-435® system.

5.2 Experimental

5.2.1 Materials

All of the chemicals were analytical grade and were used as received. Immobilized lipase B from *Candida antarctica*, Novozym-435® (Lot#047K1672 with an activity of 11,200 PLU (propyl laurate units)/g), 3-aminopropyl)triethoxysilane (3-APS or γ-APS), silica gel (Davisil®, Grade 646, pore size 150 Å, 35–60 mesh, particle size 250–500 µm, surface area 300 m^2/g, pore volume 1.15 cm^3/g, specifications provided by Aldrich Co.), glutaraldehyde and Brilliant Blue G-250 were purchased from Aldrich Co. HPLC/Spectro grade toluene and tetrahydrofuran (THF) were purchased from Tedia Co. Inc. Polyallylamine hydrochloride (PAH) and polyethylenimine (PEI), which were used for bioinspired silica synthesis, were obtained from Sigma-Aldrich. The free CALB was kindly provided to us by our collaborator Professor Richard A. Gross, Polytechnic Institute of NYU, Brooklyn, NY.

5.2.2 *Enzyme Immobilization Protocol*

The surface modification of the silica particles were performed by two different methods—namely a wet method (aqueous solution) and a dry method (toluene) [11]. In the dry method, about 1 g of porous silica gel particles were taken into a glass vial and then kept in an oven at 150 °C under vacuum (500 mmHg) for 24 h. In the wet method, the silica gel particles were transferred into a glass vial containing 10 ml of NaOH solution (pH ~ 10.6) and 100 μl of 3-APS and kept for 15 min. While, in the dry method, the silica gel particles were transferred into a glass vial containing 10 ml of toluene and 100 μl of 3-APS and kept for 15 min. Then the solutions from these glass vials were removed carefully with a micropipette and the particles were washed with 5 ml of ethanol (twice) corresponding to each of the vials. These two silica samples were labeled as wet silica and dry silica, respectively and the subsequent treatment and enzyme immobilization procedure followed was identical for both the methods. The surface coated silica gel particles with 3-APS were then cured at 150 °C for 15 min. These particles were dried in a vacuum chamber (400 mmHg) overnight. The surface modified silica gel particles were then used for the immobilization of CALB as described below.

The enzyme solution was prepared by dissolving 10 mg of CALB in 1 ml of Tris buffer solution (pH = 7.5) taken in a glass vial. About 15.6 µl of glutaraldehyde (conc. 24 %) was added to the glass vial and mixed for 2 min [19]. The solution was transferred to a glass vial containing 100 g of 3-APS coated silica gel particles (wet-silica or dry-silica) and then the glass vial was placed on a shaker at room temperature overnight. The remaining solution was carefully removed with a micropipette. The silica gel particles were washed (five times) with 1 ml of Tris buffer solution (pH = 7.5) and the washed solution was collected in five different glass vials for the analysis of protein concentration using the Bradford assay [27]. The CALB immobilized on the surface modified silica gel particles were then dried at room temperature in a vacuum chamber (400 mmHg) for 48 h and then stored in a refrigerator at 5 °C (± 1 °C) for future use. For bioinspired silica immobilised CALB, a literature method was adopted [23] except the amines used (PEI and PAH) and their concentrations were altered.

5.2.3 *Enzyme Activity Assay*

The enzyme activity was determined by the esterification of 20 mg of 1-octanol and lauric acid (1:1 mol ratio) catalyzed by 10 mg of the fresh or recovered lipase in 1 ml of isooctane at 37 °C for 1 h. The amount of product octyl laurate formed was analyzed by gas chromatography (GC).

5.2.4 *Instrumental Methods*

About 1 μl of the reaction sample was injected into the GC (Shimadzu GC-2010) column 0.5 μm × 0.25 mm × 15 m SHRX5 capillary column, split injection at 200 °C, flow rate 30.4 ml min^{-1}, FID detector 350 °C, ramp of 75–300 °C over 18 min. The residual enzyme activity was estimated from the relative peak areas of the product (octyl laurate) formed by the pure and the recovered lipases. The data were acquired and processed using the Shimadzu Class-VP software.

5.2.5 *Results and Discussion*

In the present investigation, CALB was immobilized onto 3-APS surface modified porous silica gel particles based on the optimal reaction conditions and curing temperatures that were reported in the literature [11, 28]. It was reported that the structure of 3-APS and its ability to bind on the silica surfaces is highly dependent on the pH of the treating solution [11]. The maximum number of molecules was adsorbed when the silica was treated with a solution at its natural pH 10.6. Furthermore, the number of siloxane bonds formed during the preparation of 3-APS modified silica was largest for silica gel samples with the largest amount of water at the silica gel surface [28]. The nuclear magnetic resonance (NMR) studies on a series of samples prepared by derivatization of silica gel particles with 3-APS under systematically varied silica pretreatment temperatures, reaction conditions, and post-treatment temperatures revealed that curing produced an increase in the number of siloxane bonds at the interface and was optimized in the presence of surface water at curing temperatures greater than 150 °C. Therefore, these curing conditions for the surface functionalization of silica gel particles were chosen in the present investigation. Thus, the surface functionalized silica gel particles were then used to immobilize the CALB by chemical crosslinking using glutaraldehyde [19], as discussed in the following section.

5.2.6 *CALB Immobilization on Surface Modified Porous Silica Gel Particles*

The optical micrographs of the immobilized lipase (CALB) on surface modified porous silica-gel by 3-APS are presented in Fig. 5.1 (wet silica), and in Fig. 5.2 (dry silica). There was no marked difference before and after use of the wet silica as well as the dry silica particles (Scheme 5.1).

5.2.7 *Thermal Stability of the Immobilized CALB*

In the present investigations thermal stability of CALB immobilized on 3-APS modified silica gel particles through chemical crosslinking using glutaraldehyde was studied and compared with Novozym-435®. The normalized relative enzyme

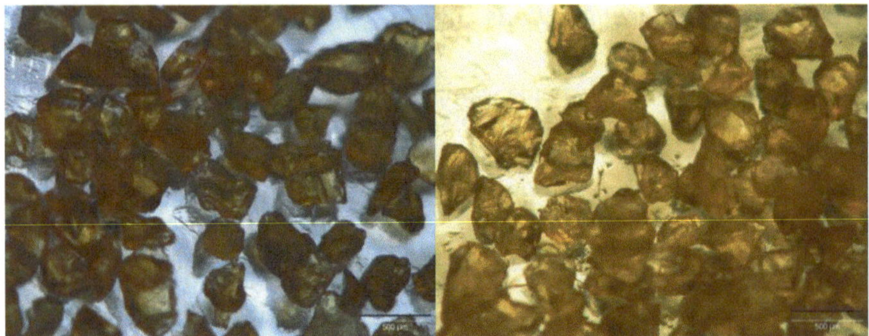

Fig. 5.1 Optical microscopy images of immobilized lipase (CALB) on surface modified porous silica gel by (3-aminopropyl)trimethoxysilane: (*left*) wet silica, and (*right*) dry silica (bar: 500 μm)

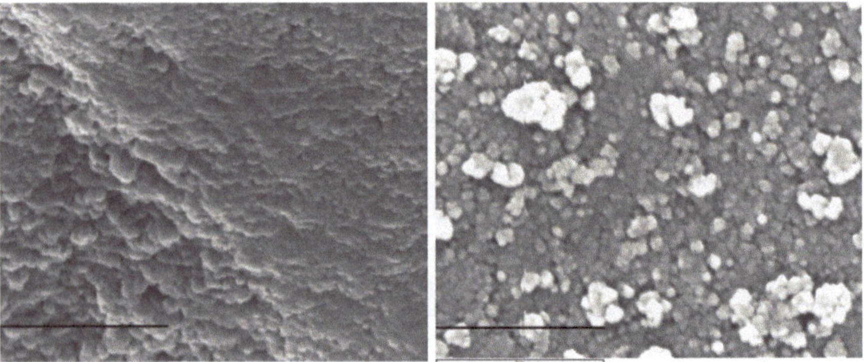

Fig. 5.2 SEM images of porous silica gel (*left*) and immobilized lipase (CALB) on surface modified porous silica gel (*right*) dry silica (bar: 1 μm)

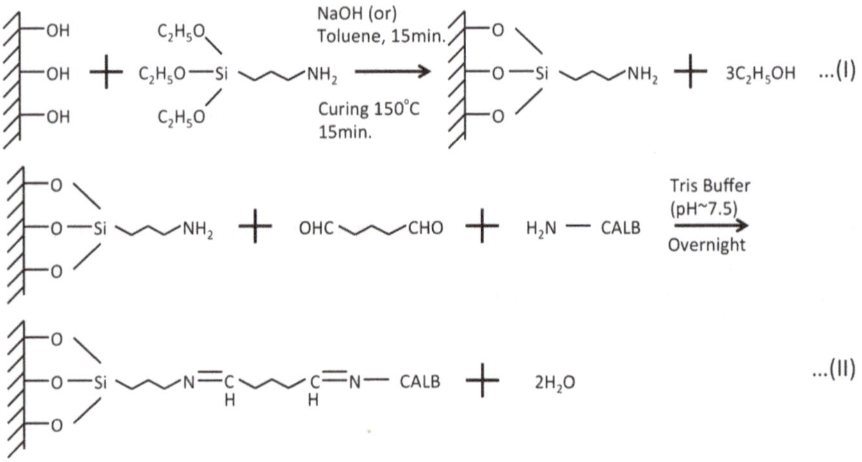

–OH
–OH + C_2H_5O, C_2H_5O–Si, C_2H_5O ~~~ NH_2 → (NaOH (or) Toluene, 15min. / Curing 150°C 15min.) –O, –O–Si, –O ~~~ NH_2 + $3C_2H_5OH$ ...(I)
–OH

–O, –O–Si, –O ~~~ NH_2 + OHC ~~~ CHO + H_2N — CALB → (Tris Buffer (pH~7.5) Overnight)

–O, –O–Si, –O ~~~ N=C(H) ~~~ C(H)=N— CALB + $2H_2O$ ...(II)

Scheme 5.1 CALB immobilization on the surface modified porous silica gel particles, **a** surface functionalization of the silica gel by 3-APS in NaOH (pH ~10.6) or in toluene followed by **b** chemical crosslinking of CALB using glutaraldehyde (pH ~7.5)

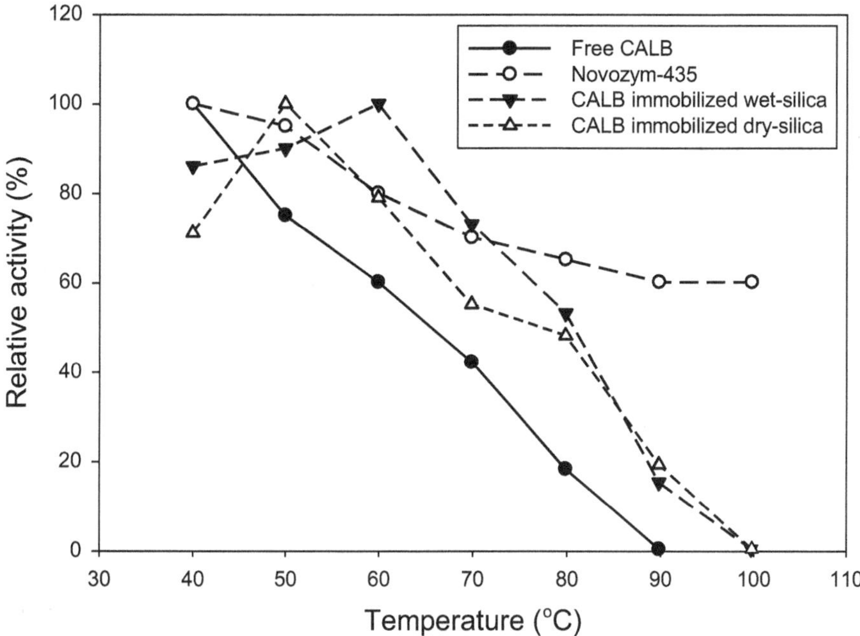

Fig. 5.3 Thermal stability (as represented by plots of the relative activity versus temperature) for the immobilized lipase B from Candida antarctica (CALB) on surface modified porous silica gel by (3-aminopropyl)trimethoxysilane compared to free CALB and immobilized CALB on an acrylic resin (Novozym-435®). The samples were incubated in toluene for 24 h at the respective temperatures and then the enzyme activities were determined using an octyl laurate assay

activities of the free CALB, CALB immobilized wet silica, dry silica and Novozym-435® incubated in toluene at different temperatures (in the range 40–100 °C) after 24 h are plotted in Fig. 5.3. The maximum activity of free CALB and Novozym-435® was found at 40 °C while it was at 50 °C for CALB immobilized dry-silica and 60 °C for CALB immobilized wet silica. The respective maximum activities were considered to be 100 % activity and the data were then normalized accordingly. The activity of the Novozym-435® incubated in toluene was found to drop from 100 % at 40 °C to 60 % at 100 °C. However, the free CALB showed no catalytic activity at 90 °C. Both CALB immobilized dry-silica and wet-silica showed similar temperature dependent activities and showed no catalytic activity at 100 °C. Furthermore, the observed drop in the activity with increasing temperature of the free CALB, CALB immobilized dry-silica and wet-silica was much steeper compared to that of Novozym-435®.

Mei et al. [29] have employed an IR microspectroscopy method to determine the distribution of enzyme in Novozym-435® and the structure of the Novozym-435®. The IR imaging showed that the CALB was localized as an external shell of the Lewatit bead with a surface thickness of 80–100 μm. The SEM analysis showed that the average pore size in the Novozym-435® beads was about 100 nm, more than 10 times larger than the size of the CALB molecule.

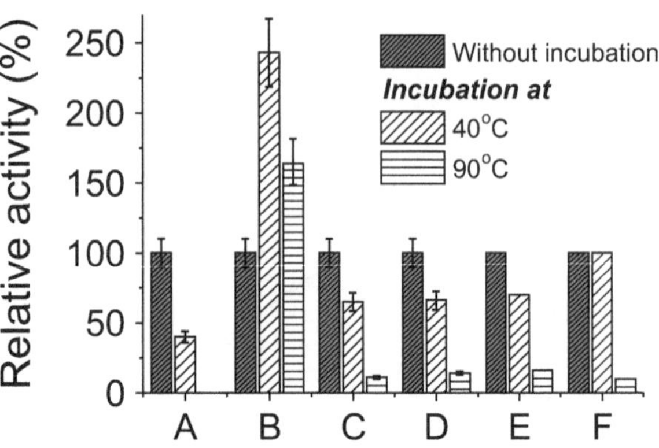

Fig. 5.4 Relative activity of free CALB (*A*), Novozym-435® (*B*), CALB immobilized wet silica (*C*), dry silica particles (*D*) and CALB entrapped in bioinspired silica (*E*: PAH-silica and *F*: PEI-silica) under without and with incubation at 40 and 90 °C in toluene for 24 h respectively. The activities were normalized and accounted for the amount of free CALB immobilized

5.2.8 Support Material Swelling

The apparent high catalytic activity of the Novozym-435®, particularly at higher temperatures, may have come from the fact that a thick lay of CALB was coated on the surface of the resin particles. It is tentatively attributed to the swelling of the resin, ~350 % by dry weight in toluene, which led to an increased surface area of the particles that, in turn, improving the access to enzymatic active sites. The swelling of the resin particles may expose new sites, which are otherwise buried deep inside the thick layer and this has resulted in an enhanced overall catalytic activity of the Novozym-435®. However, CALB immobilized dry-silica and wet-silica particles predominantly contained a monolayer of enzyme molecules on the surface of the silica particles. The chemically immobilized CALB onto the surface of the silica particles have displayed higher catalytic activity compared to Novozym-435® when the amount of CALB was taken in to consideration. The relative normalized activities of free CALB (A), Novozym-435® (B), CALB immobilized wet silica (C), CALB immobilized dry silica particles (D) and CALB entrapped in bioinspired silica (E: PAH-silica and F: PEI-silica) with incubation (24 h) and without incubation in toluene are shown in Fig. 5.4.

However, the Novozym-435® particles were found to change in appearance from opaque to translucent due to partial miscibility which led to swelling of the resin. Swelling of the resin particles up to approximately 350 % by dry weight in organic solvents was found [24]. This indicated that the resin had undergone significant morphological changes during the incubation period. For CALB immobilised in PAH-silica sample, the reduction in the enzymatic activity as a function of incubation temperature was similar to that observed for silica gel supported CALB systems. However, interestingly at 40 °C incubation, there was no reduction in enzymatic activity for CALB immobilised on PEI-silica. Although the reasons for this are not clear at this point in time, this observation is consistent with the literature [23] and is very encouraging in developing better biocatalysts in future.

5.2.9 Conclusions

Candida antarctica lipase B (CALB) was successfully immobilized onto surface modified silica gel particles. The surface modification of the silica particles was performed by two different methods—namely a wet method (wet silica in aqueous solution) and dry method (dry silica in toluene). The surface modification of porous silica gel particles was performed using (3-aminopropyl)triethoxysilane (3-APS) and then the CALB was chemically cross linked using glutaraldehyde.

A high catalytic activity of the CALB immobilized wet-silica and dry-silica particles was observed when compared to free CALB and Novozym-435. A new exciting bioinspired silica platform for enzyme immobilization was also established and it exhibited excellent candidate for developing biocatalysts. Novozym-435® was found to have superior thermal stability when compared to free CALB, to CALB immobilized wet silica and to CALB immobilized dry silica particles.

Acknowledgements We thank the National Science Foundation (NSF) for Center Funding to two of us (Clarson and Gross) under the TIE Grant NSF #0631412. SVP thanks the Royal Society for funding (Grant # TG090299).

References

1. Yamanaka S, Tanaka T (1987) Methods Enzymol 136:405
2. Manjoon A, Iborra J (1991) Biotechnol Lett 13:339
3. Chopinean J, McCafferty FD (1998) Biotechnol Bioeng 31:208
4. Gillies B, Yamazaki H (1987) Biotechnol Lett 9:709
5. Kobayashi SJ (1999) Polym Sci Part A: Polym Chem 37:3041
6. Runge M, O'Hagan D, Haufe G (2000) J Polym Sci Part A: Polym Chem 38:2004
7. Binns F, Harffey P, Roberts SM, Taylor A (1999) J Chem Soc Perkin Trans 1:2671
8. Kobayashi S, Uyama H, Namekawa S, Hayakawa H (1998) Macromolecules 31:5655
9. Matsumura S, Tsukana K, Toshima K (1997) Macromolecules 30:3122
10. Al-Azemi TF, Bisht KS (1999) Macromolecules 32:6536
11. Salis A, Meloni D, Ligas S, Casula MF, Monduzzi M, Solinas V Dumitriu E (2005) Langmuir 21:5511
12. David AE, Wang NS, Yang VC, Yang AJ (2006) J Biotechnol 125:395
13. Knezevic Z, Milosavic N, Bezbradica D, Jakovljevic Z, Prodanovic R (2006) Biochem Eng J 30:269–278
14. Ragheb AM, Brook MA, Hrynyk M (2005) Biomaterials 26:1653
15. Chen B, Hu J, Miller EM, Xie W, Cai M, Gross RA (2008) Biomacromolecules 9:463
16. Luckarift HR, Spain JC, Naik RR, Stone MO (2004) Nat Biotechnol 22:211
17. Reetz MT, Zonta A, Simpelkamp J (1996) Biotechnol Bioeng 49:527
18. Vandenberg ET, Bertilsson L, Liedberg B, Uvdal K, Erlandsson R, Elwing H, Lundstrtm I (1991) J Colloid Int Sci 147:103
19. He F, Zhuo RX, Liu LJ, Jin DB, Feng J, Wang XL (2001) Reactive Functional Polym 47:153
20. Hwang S, Lee KT, Park JW, Min BR, Haam S, Ahn IS, Jung JK (2004) Biochem Eng J 17:85
21. Dragoi B, Dumitriu E (2008) Acta Chim Slov 55:277
22. Patwardhan SV (2011) Chem Commun 47(27):7567–7582

23. Forsyth C, Patwardhan SV (2013) J Mater Chem B 1:1164
24. Poojari Y, Clarson SJ (2013) Biocat Agricult Biotechnol 2:7–11
25. Poojari Y, Beemat JS, Clarson SJ (2013) Polym Bull 70:1543–1552
26. Uppenberg J, Hansen MT, Patkar S, Jones TA (1994) Structure 2:293
27. Bradford MM (1976) Anal Biochem 72:248
28. Caravajal GS, Leyden DE, Quinting GR, Maciel GE (1988) Anal Chem 60:1776
29. Mei Y, Miller L, Gao W, Gross RA (2003) Biomacromolecules 4:70

Chapter 6
Enzymatic Modification and Polymerization of Siloxane-Containing Materials

Mark B. Frampton, Jacqueline P. Séguin and Paul M. Zelisko

6.1 Introduction

Silicones and other siloxane-derived materials are some of the most industrially and economically important polymer materials [1]. Siloxanes are valued for their thermal stability, low glass transition temperatures, resistance to oxidation, low permittivity values, and general biocompatibility [2]. The basis for these physical characteristics is due in large part to the nature of the *Si–O–Si* linkage. The *Si–O–Si* angle is rather large, approximately 145°, compared to 109.5° for typical tetrahedral systems. Given a relatively low barrier for linearization, 0.21–0.50 kcal/mol, [3–5] it has been postulated that the *Si–O–Si* link is flexible enough to range from 90–180°. In addition, the siloxane linkage is highly polarisable due to the high ionic nature the *Si–O* bond which has been determined to be approximately 51 % ionic [6]. The *Si–O* bond possesses very low rotational bond energy and is also one of the strongest chemical bonds known with covalent bond energy of 108 kcal/mol and ionic bond energy of 242 kcal/mol [2]. The increased length of *Si–O* (1.64 Å) and *Si–C* (1.87 Å) bonds compared to the *C–O* (1.41 Å) and *C–C* (1.53 Å) bonds allows for greater rotational freedom [2].

Polyesters are commercially available as a variety of products such as fibres, fillings, coatings and textiles and are typically produced under extremes of heat and pressure. Methods have been developed for producing polyester materials that include strong acids and dialkyl tin reagents. While strong acids are ideal for polyester synthesis, when acid sensitive functional groups are present, such as siloxanes, they may not be compatible due to random redistribution reactions involving the siloxane backbone or cleavage of the siloxane network [1].

Enzymatic methods, particularly those involving lipases, are becoming increasingly popular in polymer chemistry. In this regard, lipase B from *Candida antarctica*,

P. M. Zelisko (✉) · J. P. Séguin · M. B. Frampton
Department of Chemistry and Centre for Biotechnology, Brock University, 500 Glenridge Avenue, St. Catharines, ON, L2S 3A1 Canada
Tel:+1-905-688-5550 ext. 4389
e-mail: pzelisko@brockuc.a

P. M. Zelisko (ed.), *Bio-Inspired Silicon-Based Materials,* Advances in Silicon Science 5,
DOI 10.1007/978-94-017-9439-8_6

sold under the trade name Novozym-435, immobilized on a macroporous acrylic resin has been the work horse for synthesizing polymeric materials [7]. The mild reaction conditions typically associated with enzymatic catalysis makes it an ideal approach when sensitive siloxane linkages are present. Furthermore, the regio- and enantio- selectivity offered by enzymatic catalysis can be used to acquire the desired product. For example, Novozym-435 selectively esterified the C6 hydroxyl of an α,β-ethyl glucoside in the preparation of sugar-functionalized silicones [8].

Since the disclosure of the enzymatic synthesis of silicon-containing esters and amides by Dow Corning, [9] much has been learned about the relationship between siloxanes and lipases. Clarson and co-workers described the synthesis of aromatic silicone polyesters and polyamides [10, 11]. Block copolymers derived from a diacid terminated disiloxane (CPrTMDS) and polyethylene glycol, as well as triblock copolymers featuring the ring opening polymerization of ε-caprolactone have also been examined [12, 13]. Poojari et al. studied the molecular weight build up of three siloxane-polyester systems; each polyester was composed of a block derived from either 1,4-butanediol, 1,6-hexanediol or 1,8-octanediol and a block from CPrTMDS [14]. Richard Gross, in collaboration with the Scandola group, provided an enzymatic route to siloxane-containing polyester amides and compared the relative rates of polymerization of 1,8-octanediol and an aminopropyl-functionalized silicone with diethyl adipate as the acyl donor [15]. We have described the enzymatic synthesis of siloxane-containing polyesters in which all polymerizable components were siloxane-derived [16]. The number of siloxane units present in the diol monomer appeared to have little effect on the elongation kinetics, or the activation energy required for the polymerization [16].

This chapter describes our recent work, which has been focused on using enzymes to modifying siloxane-containing materials. We are primarily interested in designing environmentally benign methods for the polyesterification of siloxane-containing monomers using lipase catalysis, and to incorporate enzymatic reaction processes for synthesizing starting materials. Most of the work to be described will focus on the use of N435 (lipase B from *Candida antarctica* immobilized on a macroporous acrylic resin) as the biocatalyst.

6.2 Enzyme-Mediated Catalysis of Siloxane-Containing Materials

Siloxane-containing polyesters derived from 1,3-bis(3-carboxypropyl)-1,1,3,3-tetramethyldisiloxane dimethyl ester (CPr-TMDS-DME) and 1,3-bis(3-hydroxypropyl)-1,1,3,3-tetramethyldisiloxane (3HP-TMDS) or α,ω-bis(hydroxyalkyl)polydimethylsiloxane (HA-PDMS, Mw = 2000 g/mol by ^{29}Si NMR) have been synthesized over the temperature range of 35–160 °C using N435 as a catalyst under solvent-free reaction conditions (Scheme 6.1). A stoichiometric amount of 3HP-TMDS and CPr-TMDS-DME were combined with 5 wt% N435. Because HA-PDMS is nearly an order of magnitude more massive than 3HP-TMDS, when it was used as the

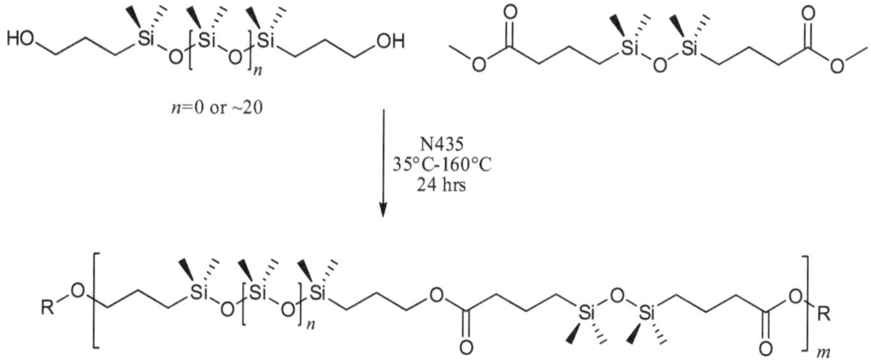

Scheme 6.1 The synthesis of siloxane-containing polyesters using enzymatic catalysis. Polyesters were produced under solvent-free conditions over 24 h time periods

acyl acceptor, the ratio of the N435 to CPr-TMDS-DME was maintained so that the catalyst loading would not change. This has not been typically done in previous enzyme-mediated polymerizations. However, due to the large mass difference in our two substrates it was prudent to control for this aspect of the reaction. The initial rate of polymer elongation and total monomer conversion were monitored by removing 10–15 μL of the reaction mixture at predetermined time intervals and performing analysis by ^{1}H nuclear magnetic resonance spectroscopy (NMR). At the end of the reaction cycle, the reaction was stopped by the addition of 5 mL of diethyl ether or chloroform when necessary. The enzyme beads were removed via filtration and the final polymer was characterized by NMR, Fourier-transform infrared spectroscopy (FT-IR), and differential scanning calorimetry (DSC).

As each N435-mediated reaction progressed the viscosity of the reaction mixture increased and the original mixture transformed from clear and colourless to slightly opaque after 24 h. Control reaction performed in the absence of any catalyst did not show any polymerization after the 24 h reaction cycle indicating that the enzyme catalyst was responsible for the polymerizations.

6.3 Structural Characterization of Siloxane-Containing Polyesters

The structural environments of the siloxane-derived polyesters were probed primarily through NMR and FT-IR spectroscopies. Typical ^{1}H NMR spectra for the disiloxane and polysiloxane polyesters are presented in Fig. 6.1. There is a triplet positioned at 4.039 ppm in the disiloxane polyester and 4.052 ppm in the silicone polyester which represents the methylene protons alpha to the oxygen atom in the newly formed ester linkage between the disiloxane-containing monomers; in the monomer this resonance is located at 3.60 ppm. There is a smaller singlet located at

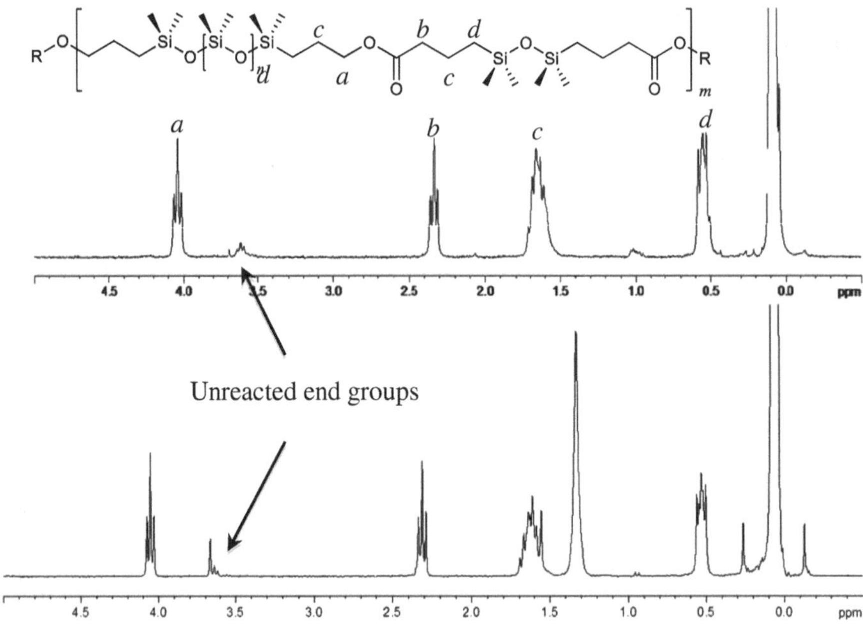

Fig. 6.1 Representative ^{1}H NMR spectra for disiloxane-polyester (*top*) and silicone-polyester (*bottom*) synthesized using N435 under solvent free conditions

3.67 ppm which can be assigned as unreacted methyl ester end groups. The methylene protons that are alpha to the carbonyl in CPr-TMDS-DME experience a small shift up-field after polymerization. The remaining multiplets, 0.54 and 1.64 ppm, represent the methylene protons that are in the α- and β- positions with respect to silicon in the monomers and final polymer; the geminal methyl groups on silicon are resonating at 0.06 ppm and remained largely unchanged throughout the polymerization process.

The ^{13}C NMR spectrum confirms the synthesis of the siloxane polyesters. The resonance for the carbonyl carbon exhibits a downfield shift and sits at 173.6 ppm while the *O*-methylene carbon can be found at 66.6 ppm; the carbon adjacent to the ester linkage was positioned at 37.7 ppm. The remaining resonances were located at 13.99 and 17.96 ppm have been identified as $CH_2\mathit{CH_2}Si$ (in monomer and polymer), and 19.06 and 22.57 ppm has been identified as the $\mathit{CH_2}CH_2Si$ environment. The ^{29}Si spectrum possessed two distinct resonances that were located at 7.30 and 7.72 ppm. These signals were in the expected range for disiloxane linkages and do not represent a possible silanol or alkoxysilane.

FT-IR group assignments were based on assignments previously reported in the literature [17]. There was a strong stretching vibration at 1737 cm^{-1} that is typical of the $\nu C=O$ stretching mode. The vibrational mode associated with the $Si\text{–}O\text{–}Si_{\mathrm{sym}}$ linkage was evident at 1050 cm^{-1}. In the polysiloxane the $Si\text{–}O\text{–}Si_{\mathrm{asym}}$ stretch was evident at 1100 cm^{-1}. The absence of peaks at 3350 and 750 cm^{-1} (*Si–OH*) suggested there was no enzymatic hydrolysis of siloxane linkages.

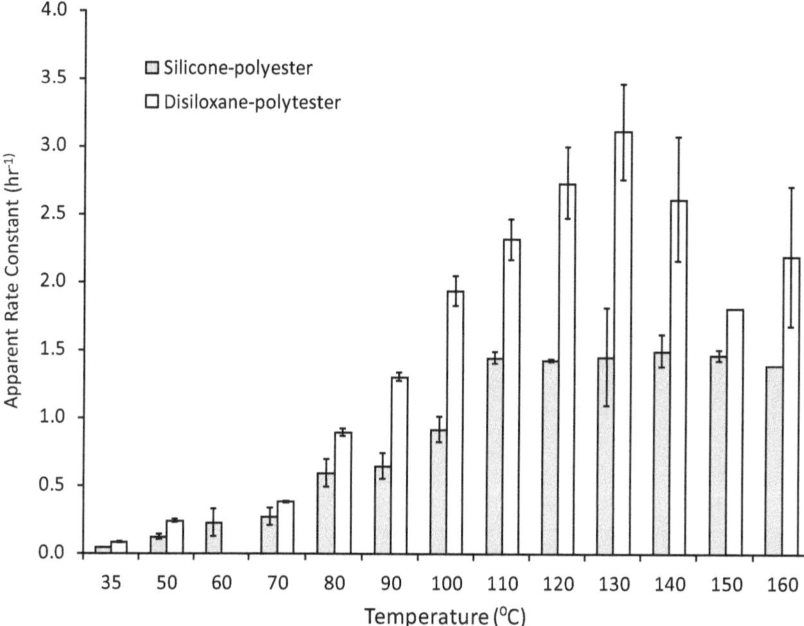

Fig. 6.2 The change in the apparent rate constant (h^{-1}) with increasing temperature. Two different acyl acceptors were used, one derived from a disiloxane (*white bars*) and the other from a polysiloxane (*grey bars*). Each bar is the average of three independent trials; error bars indicate the standard deviation

6.4 Elongation Kinetics

Polyester synthesis is a second order process [18]. The rate of polymer elongation can be determined using a plot of the average degree of polymerization (DP_{avg}) versus time, where $DP_{avg} = 1/(1-p)$ in which p is the extent of monomer conversion. The slope of the line can be taken as the apparent rate of polymer elongation. Monomer conversion was determined using the integration values from ^{1}H NMR that correspond to the *O*-methylene protons of the diol before and after esterification.

6.4.1 CPr-TMDS and 3HP-TMDS

When polyesterifications were performed in the temperature range 35–70 °C, the rate of polymer elongation was slow (Fig. 6.2). Increases in temperature to 100 °C, increased the polymerization rate to 1.94 h^{-1}. This increase is attributed to the increasing proficiency of the enzyme coupled with the ease of removal of the methanol by-product. It was hypothesized that temperatures above 100 °C would be deleterious to the functioning of the enzyme as a result of thermal denaturation. However, this was not the case (see Sect. 6.7). In fact the apparent rate constant

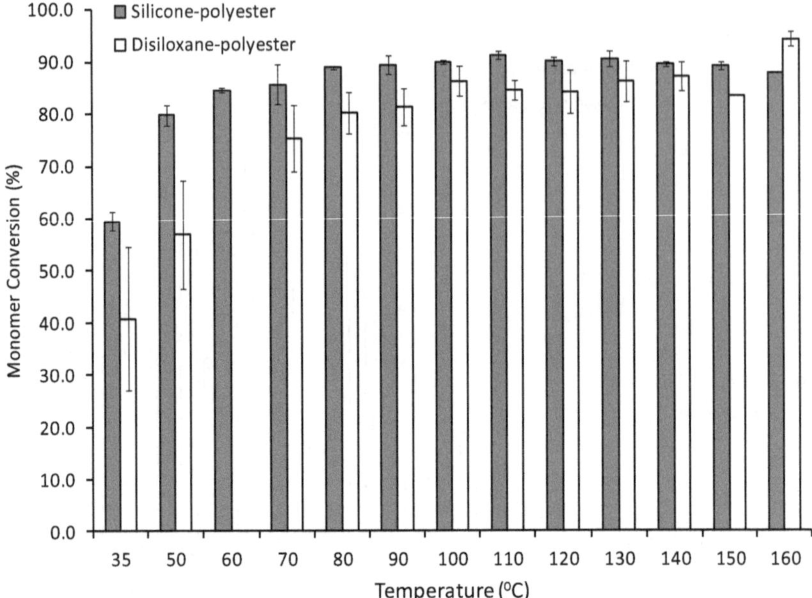

Fig. 6.3 Monomer conversion increases as the temperature is increased. Two different acyl acceptors were used, one derived from a disiloxane (*white bars*) and the other from a polysiloxane (*grey bars*). Each bar is the average of three independent trials; error bars indicate the standard deviation

associated with the polymerization continued to increase until 130 °C where the rate was 3.11 h^{-1}. Only after 130 °C did the apparent rate begin to decreases. Reactions were conducted as high as 160 °C and even at this temperature the enzyme remained active for at least a few hours. These observations suggest that the optimal polymerization temperature, when concerned with maximizing the polymerization rate and total monomer conversion, is approximately 130 °C, higher than that reported for the polymerization of poly(caprolactone) and poly(pentadecalactone) [19].

Total monomer conversion after 24 h was monitored as previously reported (*vide supra*). As the reaction temperature was increased from 35–90 °C there was a steady increase in monomer conversion (Fig. 6.3). For polymerizations conducted above 90 °C, monomer conversion seemed to plateau between 85–90 %.

6.4.2 CPr-TMDS and HA-PDMS

Many enzyme-catalyzed polymerizations, particularly those using immobilized enzymes, focus on using a fixed mass of enzyme catalyst. This is not generally problematic when the molecular weight of the monomers are close together, as in the case of increasing the number of methylene units between reactive groups, but when the mass of the repeating unit is 74 Da, in the case of a Me_2SiO– repeating unit, the ratio between the reactive end groups and the mass of the monomer can quickly become distorted. This leads to an increased amount of catalyst being used which can lead to an erroneous interpretation of the resulting kinetic data.

The mass of the polysiloxane-containing diol is nearly an order of magnitude larger than the mass of the disiloxane diol (2000 g/mol compared 250 g/mol). To ensure that a direct comparison can be made between the two systems, the stoichiometry between N435 and CPr-TMDS-DME was maintained. At lower temperatures, 35–70 °C, polyester elongation rates were generally slow, and certainly slower than the observed rates when the disiloxane diol was the acyl acceptor (Fig. 6.2). The reaction rates continued to increase as the temperature was increased reaching 1.40 h^{-1} at 110 °C. Beyond 110 °C rates did not appear to increase. Because the catalyst loading was controlled, the decrease in the elongation rates is a direct result of the increase in the polysiloxane chain length of the diol monomer. This in essence reduced the concentration of the reaction mixture reducing the number of successful binding events between the enzyme and the substrates.

Similar to reactions involving disiloxane-derived monomers, monomer conversion in the polysiloxane system increased with each successive increase in temperature (Fig. 6.3). The observed trend was similar between the two systems; that is to say at lower temperatures monomer conversion was limited by the slow rate of the reaction. However, even at 50 °C monomer conversion reached nearly 80 % after 24 h and at temperatures higher than 60 °C monomer conversion consistently reached 85–95 %. The other notable observation was that despite the slower reaction rates, monomer conversion was typically more extensive when the polysiloxane diol was the acceptor suggesting that the increased hydrophobicity had a positive effect on catalysis.

6.4.3 A Comparsion of Acyl-Donors

The increase in the molecular mass of polyesters is typically monitored by gel permeation chromatography (GPC). This method allows for the number average molecular weight (M_n), weight average molecular weight (M_w), and the polydispersity index (PDI) to be determined. Alternatively NMR-based methods can be employed to gauge the M_n of polymers. Because there are obvious differences between the methodologies that have been presented in the literature, a direct comparison between experimental conditions is not practical. To put our results into context we carried out a series of enzyme-mediated polymerizations at 100 °C using a series of dimethyl esters (fumarate, maleate, phthalate, succinate, adipate, and sebacate), in addition to CPrTMDS-DME, and a common diol, 1,8-octanediol.

The results indicated that when aliphatic diesters were the acyl donor, N435 did not distinguish between the C4, C6 or C12 esters in the polymerization of polyoctylene esters (Fig. 6.4). The olefinic diesters, dimethyl fumarate and dimethyl maleate, produced a significant drop in the reaction rate. The rate of polymerization was reduced by approximately 65 % when dimethyl fumarate was the acyl donor; with the *cis* isomer, dimethyl maleate, no polymerization was observed at all.

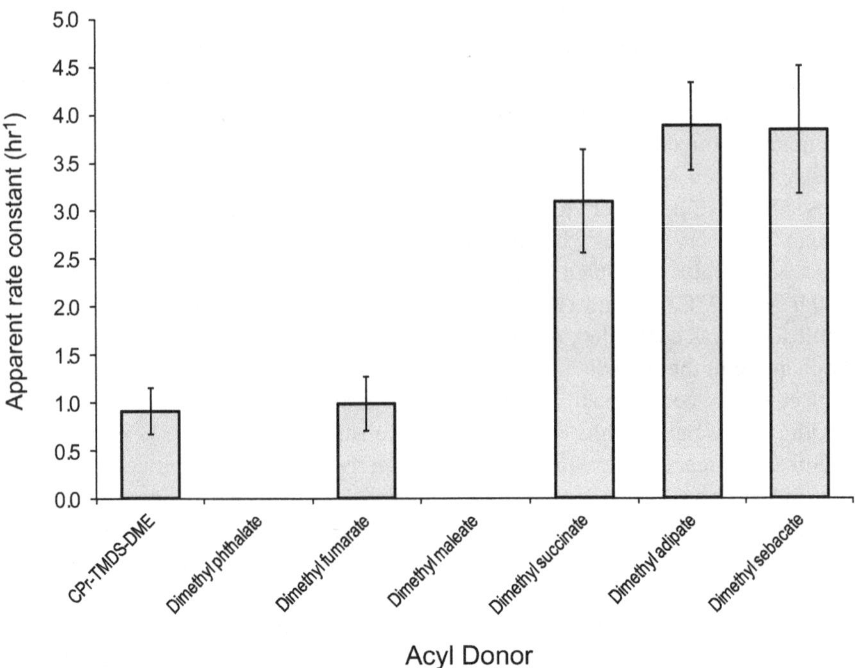

Fig. 6.4 The apparent rate constant for the N435-catalyzed polymerizations of several dimethyl esters with 1,8-octanediol at 100 °C. Each bar represents the average of three individual trial ± the standard deviation

Furthermore, dimethyl phthalate was also not a suitable acyl donor even though dimethyl terephthalate could be polymerized using N435 as a catalyst [10, 11]. These results can be attributed to the geometric arrangement of the reactive groups around the sp^2 hybridized carbons which prevent the acyl donor from binding to the active serine residue in the active site.

CPr-TMDS-DME is comparable in length to a C11 ester. Furthermore, siloxanes are some of the most hydrophobic materials that are known, which should prove to be beneficial for the functioning of N435. Generally speaking, lipases have adapted to function in hydrophobic environments. Despite this observation, N435 did not demonstrate a capacity for producing polyoctylene esters containing a siloxane block with the same proficiency as it could when charged with C4, C6 or C12 aliphatic esters. When this is taken together with the poor performance of the enzyme-catalyst with the smaller, and more geometrically constrained acyl donors, the conclusion that can be drawn is that the geometry of the acyl donor affects the proficiency with which it can be processed by N435. While CPr-TMDS-DME is not constrained by a π-system of electrons in the same manner as the fumarate, maleate, or phthalate methyl esters, the carbon-carbon single bond next to the dimethylsiloxy group is rotationally constrained as evident by the complex splitting pattern that is observed in its ^{1}H NMR spectrum. It is possible that this rotational restriction may be the limiting factor in catalysis by preventing the substrate from rearranging itself within the active site of the enzyme. Unfavourable steric interactions may also be at play. Even though CalB has a large acyl binding cleft which can accepted aromatic rings and some branched aliphatic esters, the larger size of the Me_2SiO- group may be prohibitive. The effect of steric bulk on the hydrolysis of fatty acid esters by CalB has been examined [20]. Ethyl-2-methylbutyrate was hydrolyzed with good conversion, ~90 % by GC; increasing the steric bulk on the acid side of the fatty acid ester, to ethyl benzoate and ethyl-2-phenylpropionate, elicited a decrease in conversion to between 35–45 % [20].

6.5 Thermal Properties of Disiloxane Containing Polyesters

Differential scanning calorimetry (DSC) was used to examine the thermal properties of the siloxane polyesters. Siloxane-polyester samples were transferred into aluminium pans and cooled to −150 °C at a rate of 10 °C/min. Each sample was heated at 10 °C/min to 200 °C and subsequently cooled at 10 °C/min to −150 °C. Thermal transitions were taken from a second heating scan that was done at 10 °C/min to 200 °C.

Figure 6.5 presents the DSC thermograms for siloxane monomers as well as representative siloxane-containing polyesters. The glass transition temperature (T_g) for polydimethylsiloxane (PDMS) is −125 °C. By comparison, the observed values for the disiloxane-derived diol and diester are higher, −99 and −109 °C respectively, while the T_g for the silicone-diol was more comparable at −118 °C. The T_g for the polyester synthesized from only disiloxanes was found to be −104 °C. The polyester synthesized from the silicone diol had a T_g of −115 °C, only slightly higher than the free silicone-diol. As a result of the flexibility of the disiloxane linkage and the amorphous nature of silicones, thermal transitions associated with melting and crystallization were not determinable.

6.6 A Comparison of the Activation Energy for N435-Mediated Polyesterification Reactions

The general dependence of the rate of a reaction on temperature can be quantified using the Arrhenius Eq. (6.1).

$$k = Ae^{\frac{-E_a}{RT}}. \tag{6.1}$$

The activation energy, E_a, and the Arrhenius factor can be interpreted directly from a plot of ln k vs. $1/T$ which yields a straight line with a slope of -E_a/R and an intercept

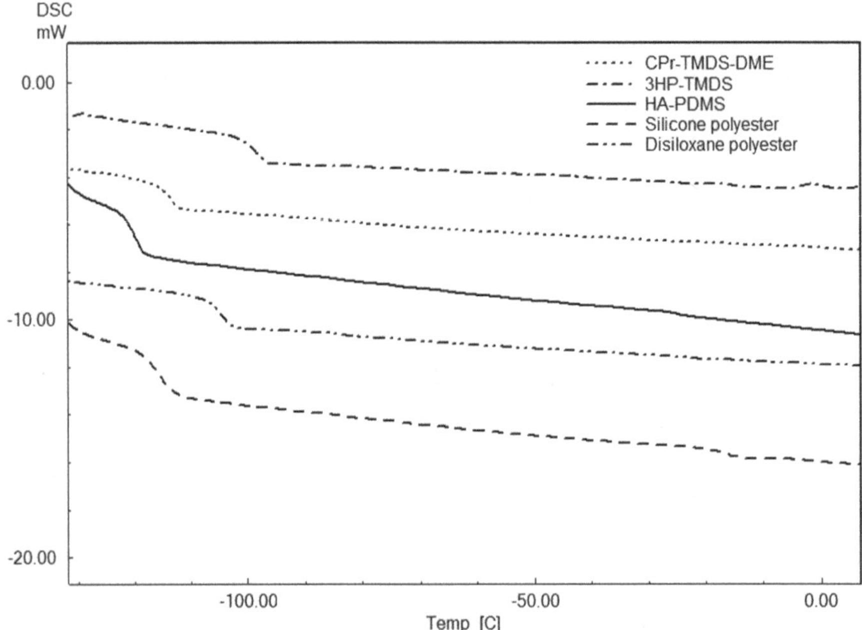

Fig. 6.5 DSC thermograms of siloxane-containing polyesters and their corresponding monomers. The thermograms presented were taken from the second heating scan that was done at a rate of 10 °C/min

of ln A (Fig. 6.6). An examination of this graph indicates two striking features. First there is an inflection point that occurs at 110 °C when the higher molecular weight diol is used. Because of this, the activation energy was determined for temperatures below this value. From the Arrhenius equation the activation energy for the silicone diol case was found to be 44.21 kJ/mol and the Arrhenius factor was $A = 1.607 \times 10^6$. When the disiloxane diol was used the activation energy was 44.06 kJ/mol and the Arrhenius factor was $A = 2.637 \times 10^6$.

6.7 Residual Activity of Novozyme-435

The residual activity of N435 was determined by recovering the enzyme containing beads and subsequently performing an assay designed to standardize for residual activity. The assay monitors the production of octyl palmitate from 1-octanol and palmitic acid. Control assays consisting of N435 that had never been used as well as an enzyme-free control reaction were included. The results of each assay were monitored using ^{1}H NMR. The change in the intensity of the resonances for the *O*-methylene protons in 1-octanol and octyl palmitate was compared. In the absence of any enzyme esterification was not detected. On the contrary, virgin enzyme

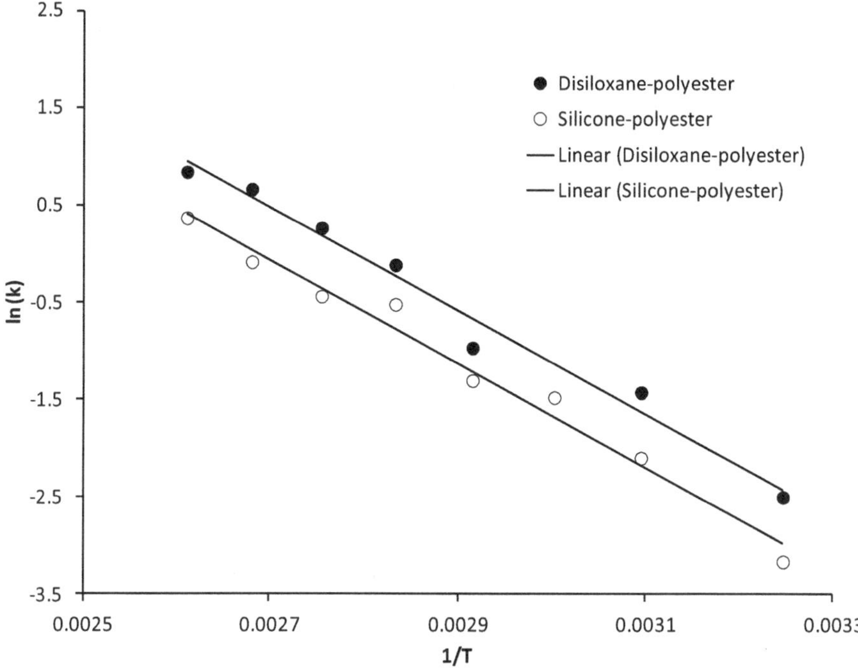

Fig. 6.6 Arrhenius plots for N435-catalysed polyesterification of siloxane-containing monomers; disiloxane diol (*hollow circles*) and polysiloxane diol (*black-filled circles*)

processed an average of 93 % of the 1-octanol within the allotted time period. Every batch that was assayed retained in excess of 95 % of its relative activity. Previous works have reported between 5 and 80 % residual activities for recovered enzymes [14, 21, 22].

While it is difficult to account for the different processing parameters that have been employed, it generally appears that hydrophobic substrates tend to favour higher residual activity. This is perhaps not surprising given the native environment in which lipases have evolved to function. Silicones are known to be some of the most hydrophobic materials, and as such disiloxane polyesters should offer a suitable environment for a lipase to function.

6.8 Thermal Tolerance and Repeated Use of Novozyme-435

In order for biocatalysis to be viable on an industrial scale the catalyst must be cost efficient, reusable and reliably perform he desired reactions with high selectivity and turnover. In light of the high residual activity of N435 after prolonged exposure to elevated temperatures it was of interest to determine the long-term thermal

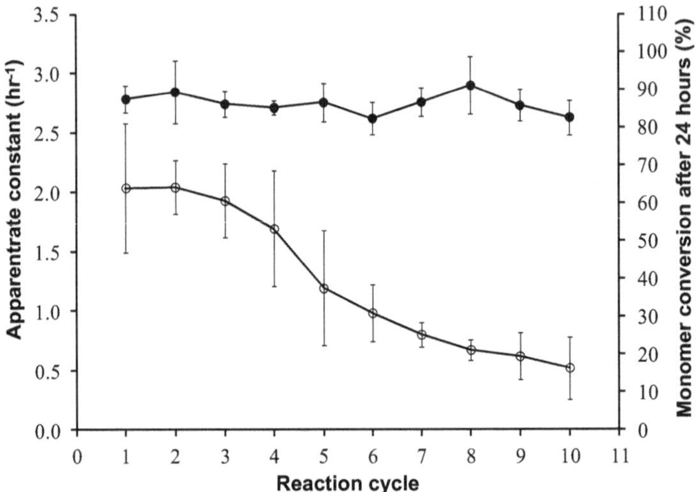

Fig. 6.7 A single batch of N435 was used for ten 24 h reaction cycles at 100 °C. The DP_{avg} (*open circles*) and monomer conversion (*black circles*) are shown

tolerance of a single batch of the immobilized enzyme catalyst. A series of 24 h reaction cycles at 100 °C were performed using a single batch of N435. The change in the reaction rate and monomer conversion was used as metrics to determine the effect of repeated long term reaction cycles. Between each consecutive use of the catalyst, the reaction mixture was cooled to room temperature and washed with diethyl ether to cleanse the acrylic beads of any residual polymer. The acrylic beads were recovered by filtering through a medium porosity glass fritted filter.

The change in the rate of polymer elongation and total monomer conversion for each of 10 successive uses of the same batch of N435 is presented in Fig. 6.7. With the exception of the first three trials in which the apparent rate constant remained fairly constant, each successive reaction cycle led to some loss in the catalytic proficiency of the enzyme. The rate decreases by approximately 50 % after six reaction cycles and by the tenth reaction cycle more than 80 % of the initial enzyme activity has been lost. Despite this loss, when the polymerizations continued for the full 24 h reaction cycle, monomer conversion reached high levels, typically in the range of 80–93 %.

These results can be compared to a previous study where a single batch of N435 was used for the ring opening polymerization (ROP) of ε-caprolactone in toluene at 70 °C over several 4 h reaction cycles [22]. N435 was shown to have lower activity in the first reaction cycle compared to subsequent reaction cycles. It was postulated that swelling of the acrylic resin allowed for any lipase on the interior of the solid support to become available to the medium and participate in the ROP of PCL.

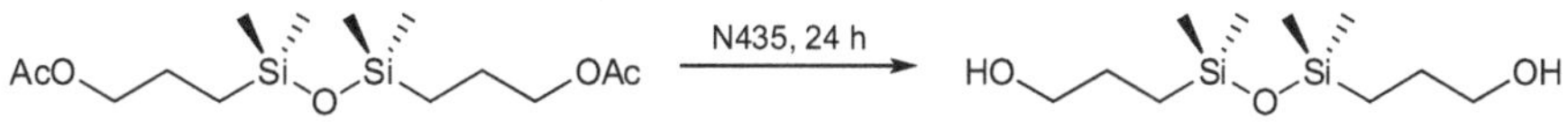

AcO–(CH2)3–Si(Me)2–O–Si(Me)2–(CH2)3–OAc —N435, 24 h→ HO–(CH2)3–Si(Me)2–O–Si(Me)2–(CH2)3–OH

Scheme 6.2 The deacylation of 3AcO-TMDS by N435 to yield 3HP-TMDS under mild reaction conditions

6.9 Enzymatic Deacylation of 1,3-Bis(3-Acetoxypropyl)-1,1,3,3-Tetramethyldisiloxane

The use of enzymes to modify organosilicon compounds has been reviewed [23, 24]. Several lipases have been used to acylate/deacylate a number of organosilicon compounds under diverse reaction conditions. Due to difficulties in obtaining some of the siloxane derived compounds that we use in pure form, we have been seeking out new synthetic routes that incorporate enzymatic steps. Recently the synthesis of 3HP-TMDS from its bis-acetate precursor (3AcO-TMDS) has been explored (Scheme 6.2). The bis-acetate is easily produced from the platinum-catalyzed hydrosilylation of allyl acetate and tetramethyldisiloxane followed by deacylation in MeOH:K_2CO_3. However, owing to the presence of the *Si–O–Si* bonding framework, yields tend to be lower than desired due to the formation of by-products which require separation, or further chemistry to remove.

Two approaches have been taken in this regard, namely methanolysis and hydrolysis *on water*, using N435 as the enzyme catalyst. The conversion of 3AcO-TMDS to 3HP-TMDS was determined by ^{1}H NMR by comparing the resonances of the *O*-methylene protons in diol (3.66 ppm) to those in the bis-acetate (4.03 ppm).

In initial experiments, a stoichiometric amount of methanol was dissolved into toluene. The enzyme catalyst was added and after 24 h of incubation, conversion was determined. Under these conditions, the optimum temperature for methanolysis was 50 °C (Fig. 6.8). Unfortunately, these conditions were not generally optimal for this process as only 47 % deacylation occurred. Elimination of toluene, by switching to an excess of neat methanol, improved conversion to in excess of 80 % at all temperatures with a maximum of 93 % at 60 °C. At 100 °C conversion dipped to a low of 80 % likely as a result of enzyme denaturation.

There are several instances where organic reaction performed on water, rather than in solvent, can be ideal. We hypothesized that because we were using a lipase, even though it does not require any interfacial activation, that deacylation under heterogeneous reaction conditions could improve on the results generated when doing deacylation in neat methanol. Enzyme free controls at deacylation via hydrolysis were performed and resulted in the recovery of the bis-acetate; no deacylation could be detected by ^{1}H NMR. The additional of N435, begining at 4 °C, promoted hydrolysis. Deacylation was minimal and reached a mere 23 %. This could be increased to 47 % by increasing the temperature to 20 °C, but hydrolysis never exceeded 82 % which was achieved at 50 °C. Temperatures exceeding 80 °C were deleterious to the enzyme catalyst and hydrolysis decreased from 80 to 65 % and finally 56 %.

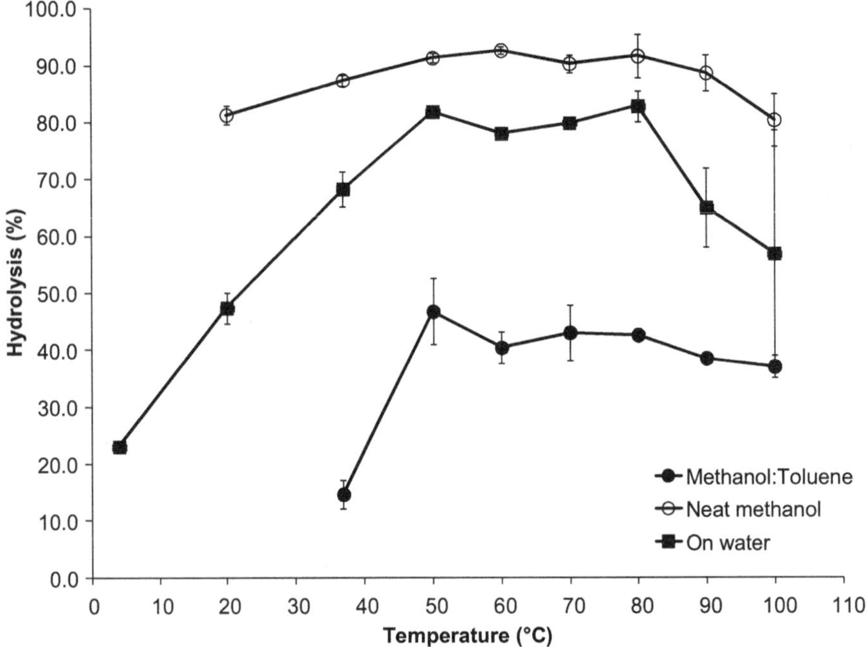

Fig. 6.8 The enzymatic hydrolysis/methanolysis of 3AcO-TMDS. Each data point is the average of triplicate trials. The standard deviation is shown

Seeking to understand the role of catalyst loading and time, two model studies were carried out to examine the effect of increasing the enzyme loading and the reaction time frame. At room temperature, hydrolysis of 3AcO-TMDS was proportional to the amount of N435 included in the reaction mixture (Fig. 6.9). The smallest amount of N435 that was used, 5 mg, afforded only 11 % hydrolysis, while increasing to 10 and 15 mg gave 24 and 33 % hydrolysis. This trend was mirrored when the time frame for the reaction was doubled to 48 h affording an increase from 11 to 22 % hydrolysis.

6.10 Conclusions

Lipases have been employed as a mild method for affecting many chemical transformations in organic and polymer chemistry. Many reports in the polymer chemistry literature have described the molecular weight build-up of the resulting polymer using N435. N435 has been shown to accept substrates of varying size, and under a diverse range of reaction conditions. A few of these reports have incorporated silicone-derived monomers to a small extent, but an extensive study of the interaction between the biocatalyst and siloxane-containing monomers has not been seen.

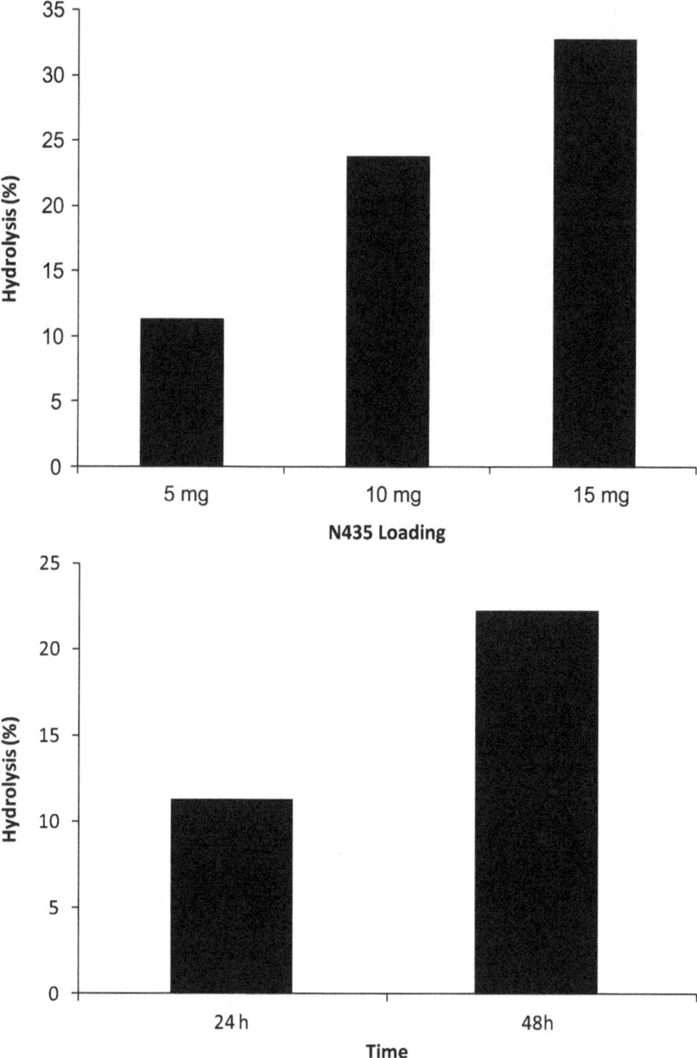

Fig. 6.9 *Top panel*: The enzymatic hydrolysis of 3AcO-TMDS. Hydrolysis is proportional to enzyme loading. *Bottom panel*: Increased reaction time leads to an increase in hydrolysis. In the absence of N435 hydrolysis is not observed

We have presented some of our work using N435 as a catalyst for preparing polyesters containing disiloxane and polysiloxane blocks, and for the preparation of organosilicon reagents containing a disiloxane framework. N435 accepted the disiloxane derived acyl donor, CPr-TMDS-DME; however when compared substrates which mimic the natural substrate for this enzyme, the disiloxane diester was not processed as quickly. Our results suggest that there lack of rotational freedom may be a limiting factor in this, but a steric contribution cannot be eliminated. While

CalB has a larger binding pocket for the carboxylic acid ester compared to some other lipases, the size of the siloxane linkage may obscure entry of the diol into the binding pocket. We are currently working of docking studies to attempt to learn the nature of the interaction between siloxane monomers and CalB.

A change in the molecular mass of the acyl acceptor, while maintaining the stoichiometry of the catalyst, resulted in a decrease in the rate of polyesterification. Dilution effects came into play as the polysiloxane diol was nearly an order of magnitude larger in mass than the disiloxane diol.

The enzymatic deacylation of a bis-acetate disiloxane has been performed. Deacylation via methanolysis in neat methanol was superior to methanolysis in toluene, or hydrolysis on water. Increasing the enzyme loading or the reaction time afforded increased conversion. Results to date have been encouraging with the achievement of >90 % deacylation. We are currently undertaking kinetic experiments to determine the time requirement for deacylation. Additionally we are continuing attempts to achieve quantitative conversion in order to eliminate the need for chromatographic separation.

Acknowledgments The authors wish to thank Dr. Thad Harroun and Drew Marquardt (Brock University) for help in acquiring DSC thermograms, Tim Jones (Brock University) for mass spectrometry assistance, and Razvan Simionescu (Brock University) for NMR assistance. Funding for this project was provided by an NSERC Engage grant to PMZ. MBF was supported by graduate scholarships from the Ontario Graduate Scholarship program, and the Queen Elizabeth II Graduate Scholarship in Science and Technology program. JPS was funded by the Brock University Experience Plus program.

References

1. Brook MA (2000) Silicon in organic, organometallic and polymer chemistry. Wiley, New York
2. White JW, Treadgold RC (1993) Organofunctional siloxanes In: Clarson SJ, Semlyen JA (eds), Siloxane polymers. Prentice Hall, Englewood Cliffs, pp 193–215
3. Gillespie RJ, Johnson SA (1997) Inorg Chem 36:3031–3039
4. Grabowski R, Hesse MF, Paulmann C, Luher P, Beckmann J (2009) Inorg Chem 48:4384–4393
5. Weinhold F, West R (2011) Organometallics 30:5815–5824
6. Eaborn C (1960) Organosilicon compounds. Butterworths Scientific Publications, London, pp 89–91
7. Gross RA, Ganesh M, Lu W (2010) Trends Biotechnol 28:435–443
8. Sahoo B, Brandstadt KF, Lane TH, Gross RA (2005) Org Lett 7:3857–3860
9. Brandstadt KF, Lane TH, Gross RA (2006) Enzyme catalyzed organosilicon esters and amides. U.S. patent 7205373
10. Poojari Y, Clarson SJ (2009) Chem Commun 6834–6835
11. Poojari Y, Clarson SJ (2010) Macromolecules 43:4616–4622
12. Poojari Y, Clarson SJ (2010) J Inorg Organomet Polym 20:46–52
13. Poojari Y, Clarson SJ (2009) Silicon 1:165–172
14. Poojari Y, Palsule AS, Clarson SJ, Gross RA (2008) Eur Polym J 44:4139–4145

15. Sharma B, Azim A, Azim H, Gross RA, Zini E, Focarete ML, Scandola, M (2007) Macromolecules 40:7919–7927
16. Frampton MB, Subczynska I, Zelisko PM (2010) Biomacromolecules 11:1818–1825
17. Lipp ED, Smith AL (1991) Chapter 11: Infrared, Raman, near-infrared, and ultraviolet spectroscopy In Smith AL (ed) The analytical chemistry of silicones. Wiley, New York, pp 305–346
18. Allcock HR, Lampe FW, Mark JE (1981) Contemporary polymer chemistry, 3rd edn. Pearson-Prentice Hall, Upper Saddle River, pp 310–324
19. Bisht KS, Henderson LA, Gross RA, Kaplan DL, Swift DL (1997) Macromolecules 30:2705
20. Naik S, Basu A, Saikia R, Madan B, Paul P, Chaterjee R, Brask J, Svendsen A (2010) J Mol Cat B: Enz 65:18–23
21. Binns F, Harffey P, Roberts SM, Taylor A (1999) J Chem Soc Perk Trans 1:2671–2676.
22. Poojari Y (2009) Enzyme immobilization and biocatalysis of polysiloxanes, Ph. D. Dissertation, University of Cincinnati, Cincinnati, Ohio, USA
23. Ryabov AD (1991) Angew Chem Int Ed Engl 30:931–941
24. Frampton MB, Zelisko PM (2009) Silicon 1:147–163

Chapter 7
Design and Thermal Properties of Interpenetrating and Intercrosslinked Biosilicate Materials

Andrew J. Vreugdenhil, Christophe Bliard, Shegufa Merchant and Suresh S. Narine

7.1 Introduction

Silane-based sol-gel chemistry has been a vibrant area of materials and coatings research over the last 40 years. This area has long been recognized as a promising and convenient means of synthesizing amorphous materials incorporating both organic and inorganic functionality. While many important early explorations of sol-gel science focused on the development of inorganic glasses using one or two silane precursors, researchers also turned to developing organically modified silicon materials (ORMOSILS).[3] The advantages of organic-inorganic hybrids were quickly recognized particularly with respect to their use in the development of flexible, tougher and more chemically diverse materials.

The organic component within the silane generally moderates the mechanical properties of the material in two possible ways, described in inorganic glass engineering as network modifiers and network formers. First, when the organic component is present as one of the four substituents on the silicon centre, it functions as a terminator or non-participating functional group in the Si–O–Si condensation

A. J. Vreugdenhil (✉) · S. S. Narine
Department of Chemistry, Trent University, 1600 West Bank Drive, Peterborough, ON K9J 7B8, Canada
e-mail: avreugdenhil@trentu.ca

Trent Centre for Biomaterials Research, Trent University, 1600 West Bank Drive, Peterborough, ON K9J 7B8, Canada

S. S. Narine
Department of Physics and Astronomy, Trent University, 1600 West Bank Drive, Peterborough, ON K9J 7B8, Canada

S. Merchant
Trent Centre for Biomaterials Research, Trent University, 1600 West Bank Drive, Peterborough, ON K9J 7B8, Canada

C. Bliard
Institut de Chimie Moléculaire de Reims, CNRS UMR7312 ICMR, Université de Reims, Champagne-Ardenne, B.P. 1039, 51687 Reims, Cedex 2, France

P. M. Zelisko (ed.), *Bio-Inspired Silicon-Based Materials,* Advances in Silicon Science 5,
DOI 10.1007/978-94-017-9439-8_7

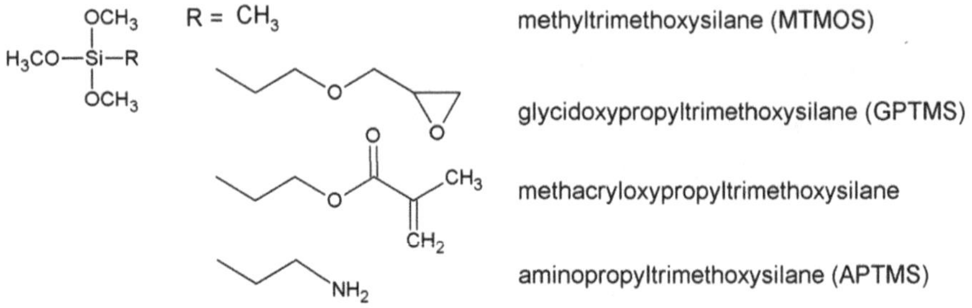

$H_3CO-Si(OCH_3)_2-R$ R = CH_3 methyltrimethoxysilane (MTMOS)

glycidoxypropyltrimethoxysilane (GPTMS)

methacryloxypropyltrimethoxysilane

aminopropyltrimethoxysilane (APTMS)

Fig. 7.1 Some network forming and modifying silane precursors

process. In this manner, the organic component permits the resulting amorphous solid to have a mechanism for relief of internal stresses during drying and densification. One of the earliest examples of this was the inclusion of methyltrimethoxysilane precursors with tetramethoxysilanes, where the methyl group is present purely as a means of reducing the condensation density of the material as it does not participate in the hydrolysis or condensation reactions forming the amorphous material.[4, 5] (Fig 7.1)

A more active role in the formation of the hybrid material, network forming, is played by the organic component when it is capable of undergoing some type of covalent bond formation to become part of the solid material network itself. This can take the form of short crosslinking interactions using organic groups on a silicon atom such as those found on alkylalkoxysilanes. Usually this approach uses a two step process where the system undergoes sol-gel hydrolysis and condensation followed by some condition which activates the organic functionality to bond either with the preformed silica network or with the organic component. Examples of this include the polymerization of acrylate functionalized silane systems where the condensation of the silica network is reinforced by the resulting polyacrylate network.[6] Alternatively, chemical crosslinking using the epoxysilane glycidoxypropyltrimethoxysilane, can be achieved following silicate condensation to generate a durable silicate network reinforced by short organic segments via acid catalyzed/thermal ring opening of the epoxide[7]or via nucleophilic attack from a crosslinking agent such as a diamine.[8, 9]

In addition to using short organic segments to form a hybrid network, longer oligomeric and polymeric organic components may be incorporated into the silicate network. As described above, these species may also function as network formers but provide opportunities for fine-tuning the extent of interaction between the organic and inorganic components. This can be achieved by controlling two important parameters: first the concentration of the organic species within the hybrid and second the extent of chemical bonding between the organic and inorganic networks within the hybrid. As with the synthesis of any molecular or macroscopic composites, the dispersal of the two components must be managed carefully in order to obtain a hybrid material that can be studied systematically.

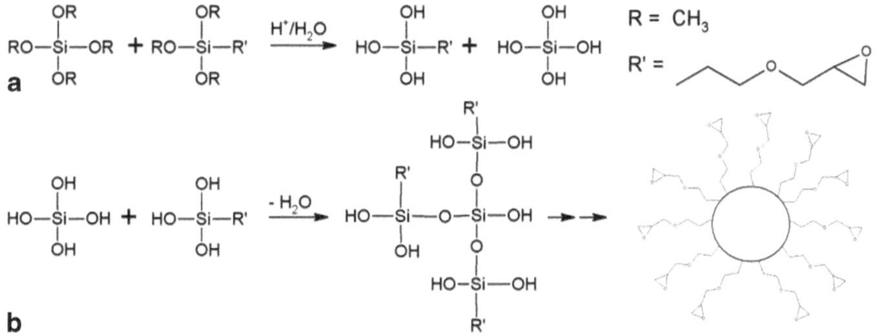

Fig. 7.2 Formation of organosilicate colloids with **a** hydrolysis and **b** condensation and aging

7.2 Biohybrid Materials

In the case of silicate based biohybrid materials, the best control is obtained if the two components are soluble in a single or a simple co-solvent system. For example, in the case of the research within our own laboratory, we have been able to achieve well behaved and robust bioorganic-inorganic hybrid formulations using sol gel techniques that produce silicate colloids with long-term aqueous stability. A typical scheme demonstrating the formation of an epoxide decorated silicate colloid is shown in Fig. 7.2. Our approach has allowed for the formation of hybrids by the combination of these preformed aqueous silicate colloids with water-soluble polysaccharides providing access to good control over the properties of the final biohybrid material.[10]

Forming a silicate colloid with a prescribed organic functionality on its surface has been shown to provide a convenient and robust method for formulating a range of interesting materials. Using a range of short chain organic crosslinking agents, we have demonstrated the utility of this approach in the formulation of several corrosion resistant coating systems,[11, 12] as a pH responsive controlled release material,[13, 14] as a means of controlling the structure in a nanodimensional multilayer device,[15] as a stabilizing host for controlling metal nanoparticle materials,[16] and in fabricating a passive sampling device for metal ions in the environment.[17]

The initial idea behind these materials was to combine the durability of a silicate material with the convenient reactivity of epoxy-amine organic reactions. However, in addition to the utility of this approach, our work has also demonstrated that the small organic crosslinking agents were responsible for significant aspects of the physical and chemical properties of the resulting hybrid. In particular, the individual strength of the nucleophile and the crosslinker size contributed to the extent of epoxy-amine crosslinking and thus the degree of interaction between the hybrid material components.[11] Based on these findings and their interesting contribution to the properties of the resulting hybrid material, we were motivated to explore hybrids based on much larger organic species, particularly ones which would provide us with access to a range of extent of interactions between the inorganic and organic components. The organic component, which could provide us with both the size and flexible reactivity that we desired, was a simple, partially-hydrolysed polysaccharide, soluble starch.

The widely used commercial product known as soluble starch is typically a corn starch which has been processed to reduce its molecular weight to enable it to be more readily dissolved in aqueous solution. Although not extensively soluble in water, solutions of 1 to 3 % by weight can be prepared. At the higher end of these concentrations, the starch solutions show a significant increase in viscosity relative to that of water and a resulting reduction in processability for forming hybrid materials using sol-gel techniques. However, materials based on a solutions containing less than 3 % soluble starch demonstrated a significant range of interesting properties.

Using soluble starch as the organic component in our biohybrid materials provided us with the ability to explore different degrees of network interaction between the inorganic and organic components. The conceptually simplest situation is one in which an interpenetrating but not covalently interacting network is formed. The interpenetrating networks (IPN) are well described in the formation of various types of hybrid materials. In general, one polymeric network is dissolved or swelled to allow for the formation of the second network *in situ.* For example, a review by Mauritz et al. from 2007 describes research using the organic component as a scaffold for inorganic sol-gel polymerization with Nafion as the scaffold and a range of silane and other metal alkoxides as the inorganic precursors.[18] Similarly, use of the readily water soluble tetrakis(2-hydroxyethyl) orthosilicate (THEOS) silane has facilitated the development of interpenetrating hybrid materials based on water soluble biopolymers. The use of THEOS in combination with polysaccharides was reviewed in 2005 [19]describing the potential that these developments have for improving the range of possible applications for biohybrids. An example of such an application was described by Wang et al. where chitosan has been used as the organic component, with THEOS undergoing *in-situ* hydrolysis and condensation to form an interpenetrating biosilicate hybrid for use in an amperometric biosensor for hydrogen peroxide.[20]

The availability and wide range of pre-existing biopolymers such as chitosan, starch and carrageenan along with the diversity of monomeric silanes tends to ensure that typically the biopolymer is used as the high molecular weight species while the silane precursor undergoes condensation to develop the inorganic network in the presence of the biopolymer. In our own research we have instead developed a process where the silicate network is also preformed which allows us to use the resulting silicate nanoparticles as material building blocks regulating the degree of interaction with the biopolymer component through the use of organic crosslinking interactions such as epoxy-amine crosslinking.

We have found that by working with both a preformed biopolymer and preformed silicate nanoparticle, the material properties of the resulting hybrid can be more systematically evaluated as the concentration, presence or absence of either of the two components does not alter the chemical conditions of the formation of the individual components. Rather, the two components are prepared separately in a

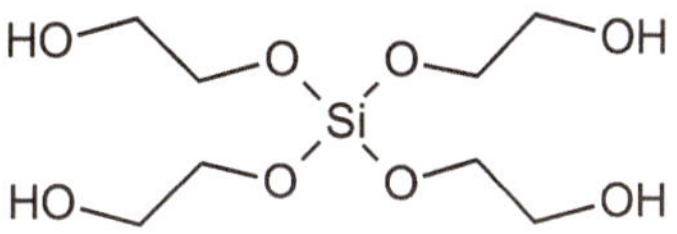

Fig. 7.3 The structure of tetrakis(2-hydroxyethyl) orthosilicate (THEOS)

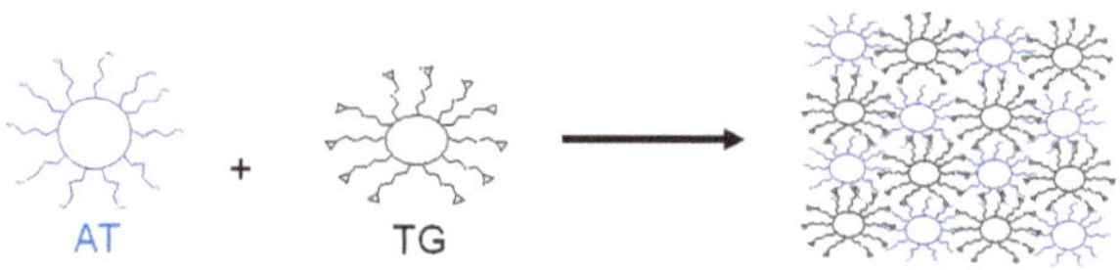

Fig. 7.4 Amine and epoxide silicate colloids forming an organosilicate material

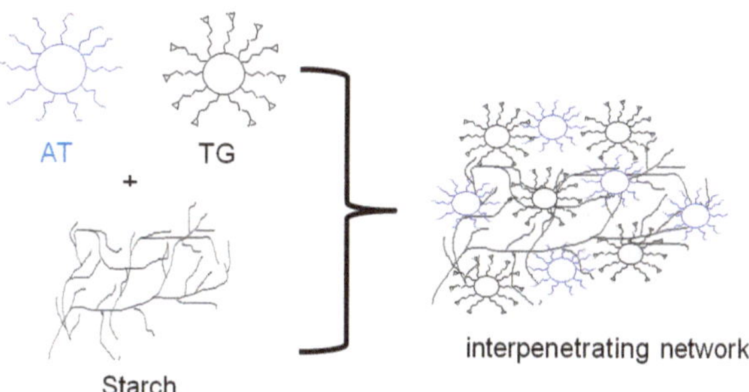

Fig. 7.5 Epoxy-amino silicate starch interpenetrating network

reproducible manner followed by formulation of the biohybrid composite where the effect of the degree of interaction between the two components can be investigated.

A demonstration of this approach is our synthesis of a series of biohybrid silicate materials as illustrated in Fig. 7.3 through Fig. 7.6. We selected a silicate material made from two different silicate nanoparticles. This allowed us to generate two organosilicate reaction mixtures containing epoxide decorated silicate nanoparticles (TG) and amine decorated silicate nanoparticles (AT). Individually, these silicate nanoparticles are stable in aqueous solution over a period of weeks. However, when combined, they undergo rapid epoxy-amine crosslinking to form a glassy material in approximately 30 min. These interactions are shown schematically in Fig. 7.4.

To these epoxy and amine decorated preformed silicates, we have added starch with and without amine functionalization. In the absence of an amine functionalization of the starch, the use of a 1–3 % solution of starch in water results in a readily formed interpenetrating network with the silicate network and the starch as preformed, solution borne polymers. The solidification of the biohybrid material occurs due to the rapid epoxy-amine crosslinking of the silicate. This is shown schematically in Fig. 7.5. The physical, optical and mechanical properties of these hybrids are significantly different from those observed for the starch free materials even for materials made with 1 % solutions of starch.

The range of properties for biohybrids constructed based on these silicate starting materials have been further extended by using a biopolymer that will interact covalently with the epoxy or amine decorated silicates. This can be done readily by

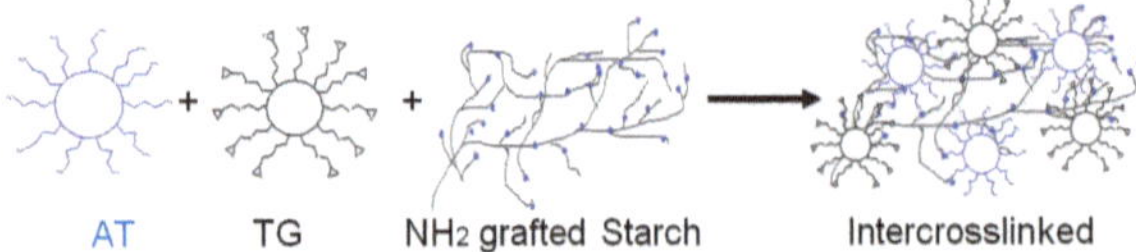

Fig. 7.6 Intercrosslinked biohybrid from amino-starch and epoxy-amine silicate colloids

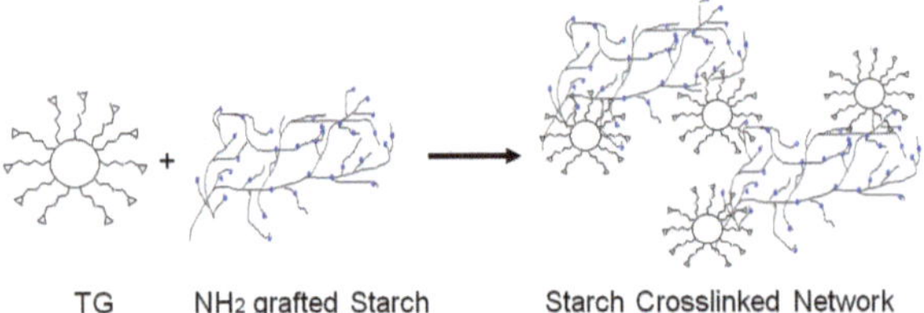

Fig. 7.7 Biohybrid formation using epoxysilicate colloids and amine functionalized starch

functionalizing starch in aqueous solution via a one-pot reaction using epichlorohydrin under basic conditions followed by reaction with ammonia to yield propyl amine groups distributed over the length of the starch backbone. Work by Ayoub et al. has demonstrated that the reaction performed with glycidyl ammonium analogs in the presence of a moderate excess of NaOH results in degrees of substitution (DS) in the 0.05 range, or one substitution every 20 saccharide rings approximately. [21, 22] Reacting an alkaline starch solution with epichlorohydrin in similar conditions, then treating the resulting glycidyl starch product with excess ammonia leads to a 0.085 DS aminoglyceryl grafted starch. This corresponds to one substitution every 11 or 12 sugar units as measured by elemental analysis and NMR. The reaction can be followed readily in the solid state by infrared spectroscopy to demonstrate the appearance of the epoxide CH_2 stretch at 3045 cm^{-1} and its subsequent consumption.

By reacting this amine functionalized starch with the prefabricated epoxy and amine silicates, we have formed intercrosslinked polymer networks with covalent interactions between the epoxy-silicate and the amine functionalized starch as well as between the epoxy and amine silicates as above in the IPN biohybrid. The intercrosslinked polymer is shown schematically in Fig. 7.6.

A different intercrosslinked material can also be obtained by excluding the amine decorated silicate so that only the epoxy silicate and the amine functionalized starch are used to produce a solid. This is shown schematically in Fig. 7.7. This material is typically much slower to solidify as the concentration of amine groups is relatively low however the resulting material is much more dense and glassy. This is thought to be due to the fact that more of the amine groups of the starch are reacted with the silicate epoxide whereas in the case of the intercrosslinked hybrid using both amino silicate colloids and amine functionalized starch, the more rapidly reacting amino silicate colloids are predominantly responsible for crosslinking the material and more of the amine groups on the starch remain unreacted, allowing for greater flexibility in the material.

7.3 Biohybrid Material Thermal Properties

Inclusion of a biopolymeric component within a silicate based matrix can be responsible for varying a large number of the properties of the resulting hybird. In our work, we have used starch inclusion and in particular the type of interaction between the starch and silicate components to change the mechanical and thermal properties of the biohybrid. Using thermal gravimetric analysis (TGA), we have been able to explore the effect that the type of interaction has on the water retention and thermal degradation of the individual components.

The results are shown individually in Fig. 7.8 through Fig. 7.12 for the silicate colloid materials itself, starch itself and then the three types of interactions between the two components: interpenetrating, intercrosslinking and intercrosslinking with only amine functionalized starch. The TGA results for the biohybrid materials each show three main regions of interest. First, the water loss peak in each case occurs around 100 °C. Following the loss of water, the materials are stable over the next 200 °C until the appearance of the thermal decomposition peaks corresponding to the destruction of the starch at 300 °C for pure starch and the alkyl chains in the silicate colloid above 375 °C. (Fig. 7.9, Fig. 7.10, Fig. 7.11)

The water loss peak identifies some clear differences in these materials. In the case of epoxy-amino silicate colloid material without any starch, the water loss peak is much sharper than the thermal decomposition peaks and represents a small percentage of the total mass loss whereas the water loss peak in all of the starch

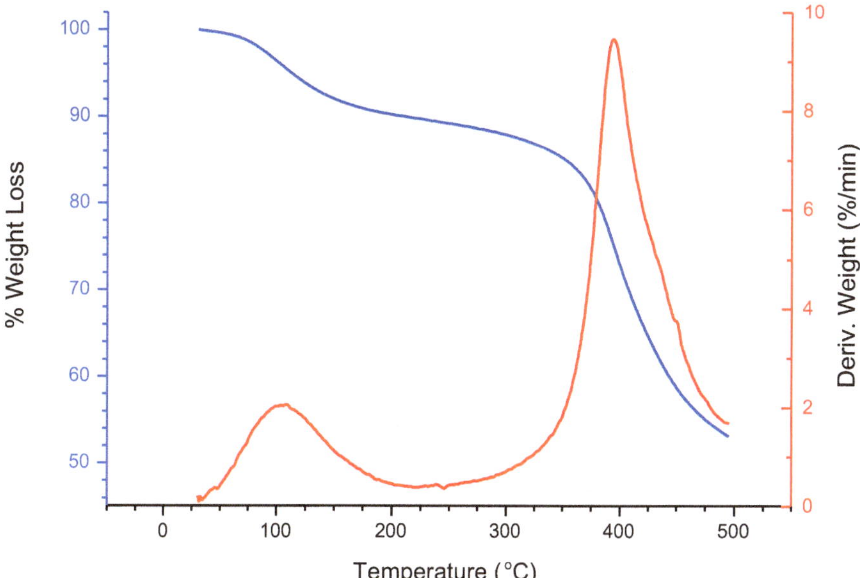

Fig. 7.8 TGAT epoxy-amino silicate material TGA

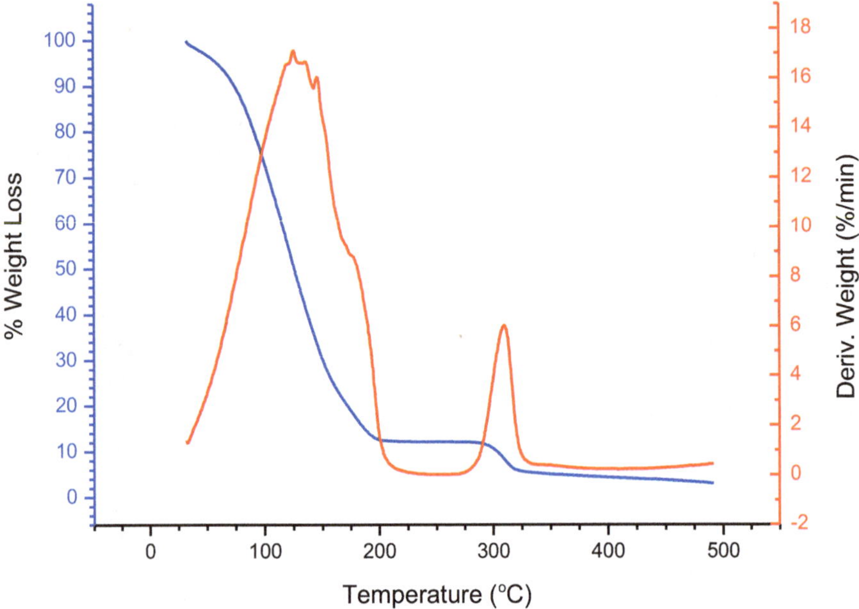

Fig. 7.9 Soluble starch TGA

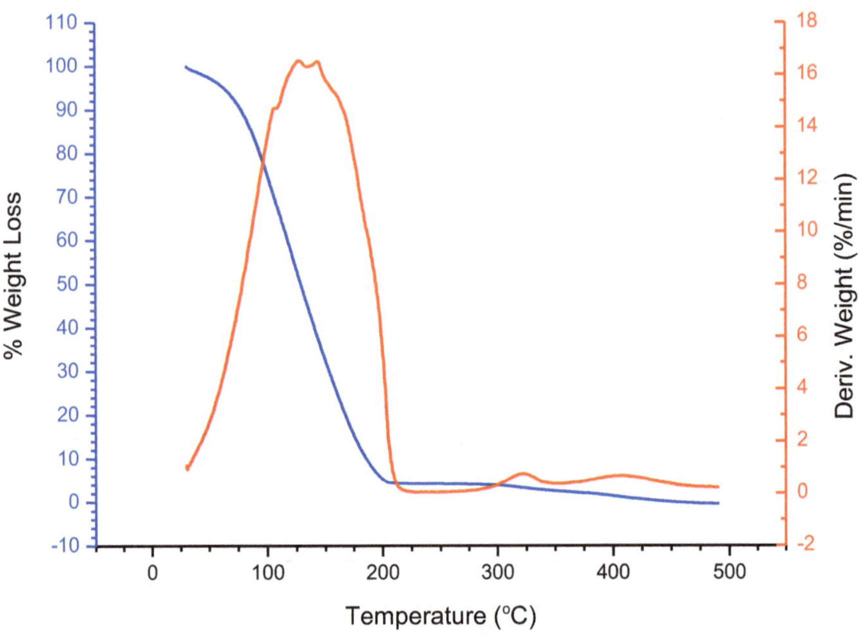

Fig. 7.10 Starch—epoxyamine silicate colloid IPN biohybrid TGA

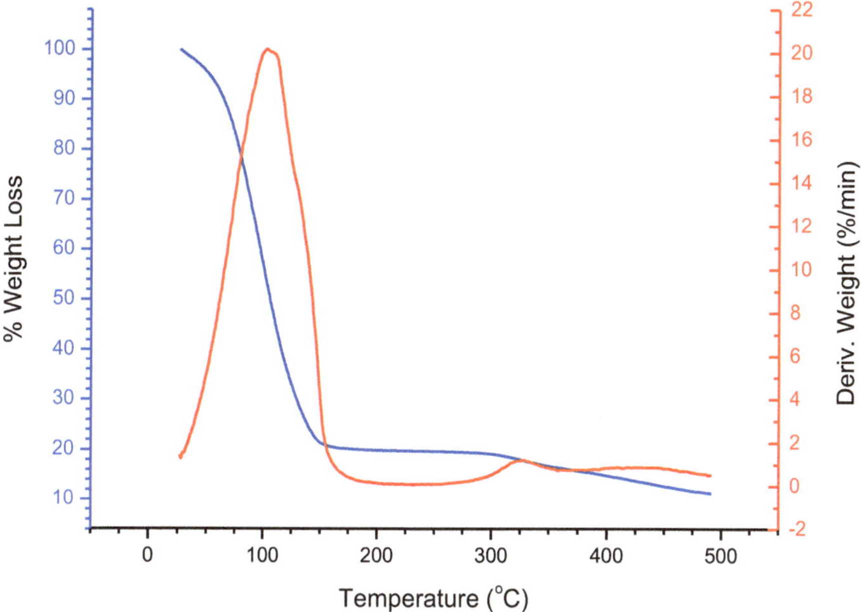

Fig. 7.11 Amino-starch epoxyamine silicate intercrosslinked biohybrid TGA

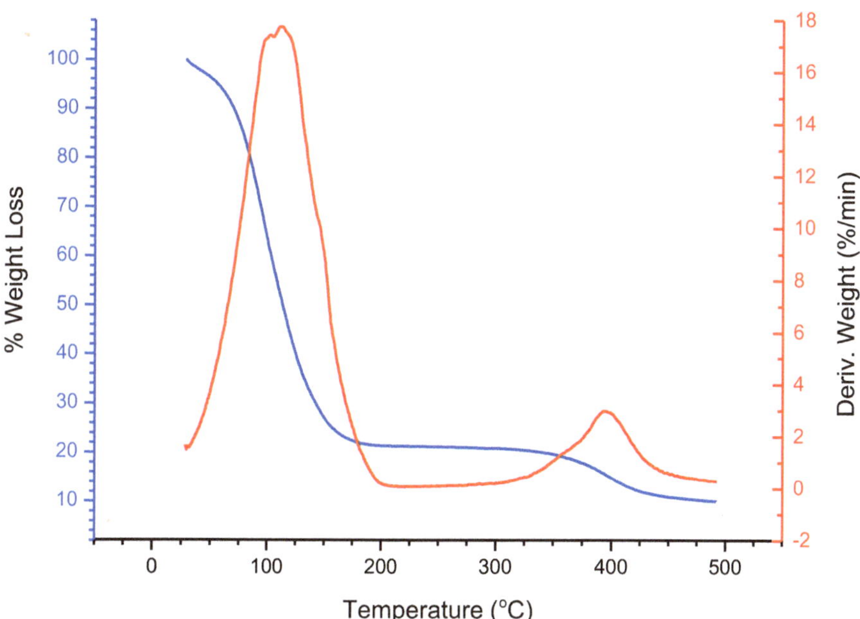

Fig. 7.12 Amino starch crosslinked epoxy silicate biohybrid TGA

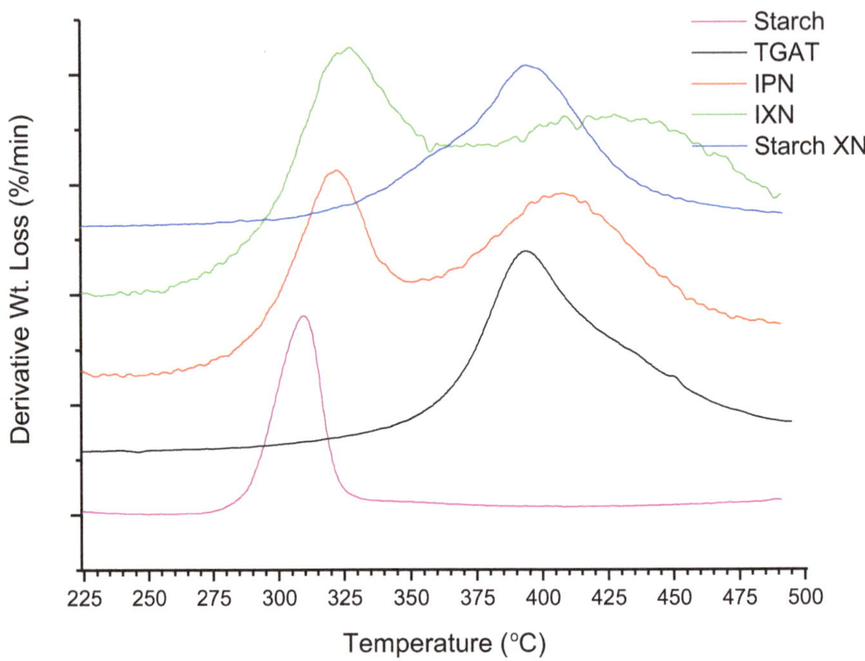

Fig. 7.13 Derivative weight loss of the decomposition peaks of the five biohybrid materials

containing materials is the major feature and demonstrating the large difference the presence of even a small amount of starch makes to these materials in terms of their ability to store water. In addition the interaction between water and starch in these biohybrids varies as well. For pure starch the water release extends from 50 °C to just above 200 °C. This is similar to the behaviour observed for the IPN of the biohybrid as the starch networks continues to closely resemble that of pure starch. However the water release observed for the intercrosslinked hybrids is significantly narrower with completion of the water loss below 180 °C and much of that occurring at temperatures around 100 °C. This behaviour seems to be dependent on the extent of interaction between the starch component of the hybrid material and the degree to which the interaction disrupts the native structure of the starch.

The second major feature of the thermal analysis of the hybrid materials occurs between 300 and 450 °C depending on the hybrid. The derivative of the weight loss of the TGA results for this region is shown for all of the materials in Fig. 7.13. These weight loss peaks correspond to the decomposition of the organic components of the materials. In the case of pure starch a single narrow feature in the derivative of the weight loss is observed from 290 °C and 320 °C with a peak 310 °C due to the decomposition of the starch. For the organosilicate material without starch, no decomposition is observed until 360 °C with a broad peak at 395 °C tailing off to 470 °C. This delayed decomposition is due to the short propyl-chains present in the organosilica materials. In the case of the IPN, the decomposition for both the starch and organosilica are shifted to slightly higher temperatures however the starch and organosilica largely behave as separate components in this minimally interacting hybrid material. Finally the two intercrosslinked hybrid materials show the most dramatic changes in their thermal behaviour. In the case of the intercrosslinked hybrid involving both aminosilica and aminostarch crosslinking with the epoxide, the starch decomposition is shifted to 327 °C while the propyl chain degradation of the organosilica component is broadened and shifted considerably to 431 °C. We expect that this is due to the increased number of thermal relaxation avenues available for the organosilica component through its close interaction with the larger organic network. Finally, the starch crosslinked biohybrid demonstrates the most substantial change in its thermal decomposition behaviour. The starch component and the organosilica component have almost entirely coalesced with a low temperature shoulder visible at 365 °C and a peak at 395 °C. This represents a significant thermal stabilization of the starch component of between 70 to 85 °C above that of native starch. The coalescence of the two thermal features is consistent with the extensive chemical interaction found in this biohybrid where the starch is behaving as the only crosslinking agent for the organosilica component. This significant change in the decomposition temperature of the starch is consistent with the behaviour observed for the water loss temperature in this starch-crosslinked hybrid material given the extent of modification of the starch morphology in this hybrid.

7.4 Conclusions

Biosilicate hybrid materials using starch components are a convenient and effective approach to constructing durable materials while incorporating polysaccharides derived from renewable resources. In our work we have focused on the combination of biopolymers and preformed organosilicate colloids synthesized using sol-gel techniques. With these convenient and robust building blocks, we have been able to fabricate a range of biohybrid materials with varying degrees of interaction between the components. The different chemical interactions between the components results in significant changes to the physical properties of the materials as demonstrated in our work by the changes in the water loss and thermal decomposition of the hybrids.

References

1. Gill I, Ballesteros A (1998) J Am Chem Soc 120:8587–8598
2. Silva SS, Ferreira RAS, Fu LS, Carlos LD, Mano JF, Reis RL, Rocha J (2005) J Mater Chem 15:3952–3961
3. Brinker CJ, Scherer GW (1990) Sol-Gel science: the physics and chemistry of sol-gel processing. Academic, New York
4. Schmidt H, Seiferling B (1986) Mater Res Soc Symp Proc 73:739–750

5. Li X, King TA (1996) J Non-Crystal Solids 204:235–242
6. Sarmento VHV, Frigerio MR, Dahmouche K, Pulcinelli SH, Santilli CV (2006) J Sol-Gel Sci Technol 37:179–184
7. Metroke TL, Kachurina O, Knobbe ET (2002) Prog Org Coat 44:185–199
8. Vreugdenhil AJ, Balbyshev VN, Donley MS (2001) J Coatings Tech 73:35–41
9. Gizdavic-Nikolaidis MR, Zujovic ZD, Edmonds NR, Bolt CJ, Easteal AJ (2007) J Non-Cryst Solids 353:1598–1605
10. Vreugdenhil AJ, Bliard C (2010) "Inter-cross-linking siloxane polysaccharide hybrid materials: formation, stability, and sequestering behavior", international chemical congress of pacific basin societies, Honolulu, HI, United States, December 15–20 MACRO–7
11. Vreugdenhil AJ, Gelling VJ, Woods ME, Schmelz JR, Enderson BP (2008) Thin Solid Films 517:538–543
12. Croes KJ, Vreugdenhil AJ, Yan M, Singleton TA, Boraas S, Gelling VJ (2011) Electro Chimica Acta 56:7796–7804
13. Vreugdenhil AJ, Woods ME (2005) Prog Org Coat 53:119–125
14. Woods ME, Vreugdenhil AJ (2006) J Mater Sci 41:7545–7554
15. Vreugdenhil AJ, Horton JH, Woods ME (2009) J Non-Cryst Solids 355:1206–1211
16. Vreugdenhil AJ, Pilatzke KK, Parnis JM (2006) J Non-Cryst Solids 352:3879–3886
17. Zhou Y, Stotesbury T, Hintelmann H, Vreugdenhil AJ, Dimock B (2013) Chemosphere 90:323–328
18. Mauritz KA, Hassan MK (2007) Polymer Rev 47:543–565
19. Shchipunov YA, Karpenko TY, Krekoten AV (2005) Compos Interfaces 11:587–607
20. Wang GH, Zhang LM (2006) J Phys Chem B 110:24864–24868
21. Ayoub A, Bliard C (2003) Starch/Stärke 55:297–303
22. Ayoub A, Berzin F, Tighzert L, Bliard C (2004) Starch/Stärke 56:513–519

Chapter 8
Bioactive Amino Acids, Peptides and Peptidomimetics Containing Silicon

Scott McN. Sieburth

8.1 Silicon in Amino Acids and Peptides

More than 50 years have passed since the first silicon-substituted amino acid was prepared by Birkofer and Ritter, β-trimethylsilyl alanine **1**, Fig. 8.1 [1]. Many other silicon-containing amino acids have been prepared during the intervening years and these have been broadly investigated [2]. Among these interesting non-natural amino acids, an intestively studied analog is sila-proline **2** [3]. In contrast, the elusive amino acid **3** has only been prepared as a derivative of the parent structure and its limited stability has precluded extensive investigation [4]. Substitution on nitrogen, as in **4**, allows reactions to be performed on peptides [5–7]. When silicon is part of a peptide backbone, silanediol **5**, it can mimic a hydrated dipeptide and these have been studied as inhibitors of protease enzymes. This chapter will focus on structures related to **1–3** and **5**, including their uses and their syntheses.

8.2 Where Silicon Can—and Cannot—be Used

The introduction of silicon into biologically active molecules has been pursued in various guises for decades and been the subject of several reviews [8–17]. There are no naturally occurring organosilanes. Silane groups can be attached to bioactive molecules, or silicon can replace carbon within the bioactive structure. Several principles are fundamental to the investigation of organosilanes in a biological context: the oxidative lability of the Si–H bond makes them unsuitable, [18] unsaturation proximal to silicon can lead to instability, [19] 3- and 4-membered rings containing silicon are very strained, [20–22] multiple bonds to silicon are extremely reactive

S. McN. Sieburth (✉)
Department of Chemistry, Temple University, 1901 N. 13th Street, Philadelphia, PA 19122, USA
e-mail: scott.sieburth@temple.edu

P. M. Zelisko (ed.), *Bio-Inspired Silicon-Based Materials,* Advances in Silicon Science 5,
DOI 10.1007/978-94-017-9439-8_8

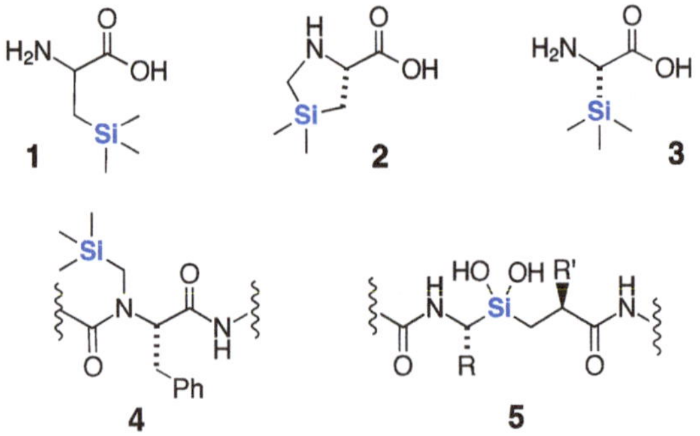

Fig. 8.1 Silicon-containing amino acids and peptides

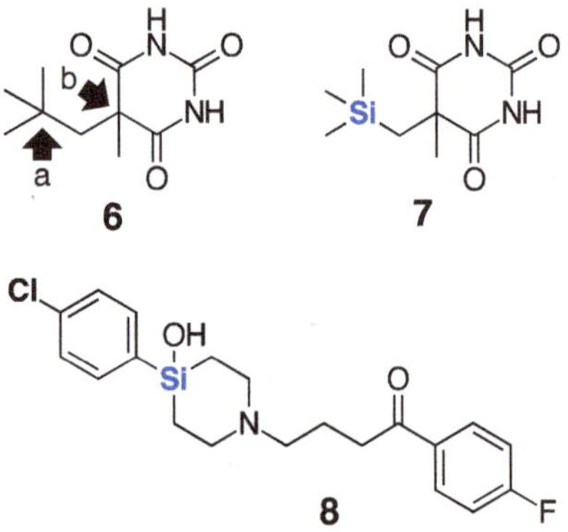

Fig. 8.2 Examples of replacement of tetrasubstituted carbon by silicon

[23] and Si–heteroatom bonds are easily hydrolyzed. Taken together, these principles seemingly identify quaternary carbons as the only carbons that can be readily replaced by silicon in the design of bioactive organosilanes.

The application of these principles to a Si/C swap are exemplified in Fig. 8.2. For the seven unique carbons in known barbiturate **6**, [24] only the quaternary carbons (arrows) are suitable for replacement by silicon. Replacement of the neopentyl carbon with silicon leads to **7** (arrow a), which is notably faster acting and has a longer effect than the non-silane **6** [25]. Replacement of the quaternary carbon between the carbonyl groups in **6** (arrow b), has not been reported and structures with silicon between two amide carbonyls are unknown. Silicon close to carbonyl groups tend to have limited stability [26].

Replacement of carbon with silicon is not, however, restricted solely to quaternary carbons. Haloperidol is a dopamine receptor antagonist incorporating a tertiary alcohol. Replacement of the tertiary alcohol carbon with silicon yields sila-haloperidol **8** [27, 28]. Once again, the only site in haloperidol that is suitable for substitution of silicon for carbon is at the position shown. Sila-haloperidol **8** has very interesting pharmaceutical properties [27]. Metabolism studies of **8** have provided much insight into comparative metabolic fates for silanes relative to their carbon analogs [29].

Fig. 8.3 α-Silyl amino acids and derivatives

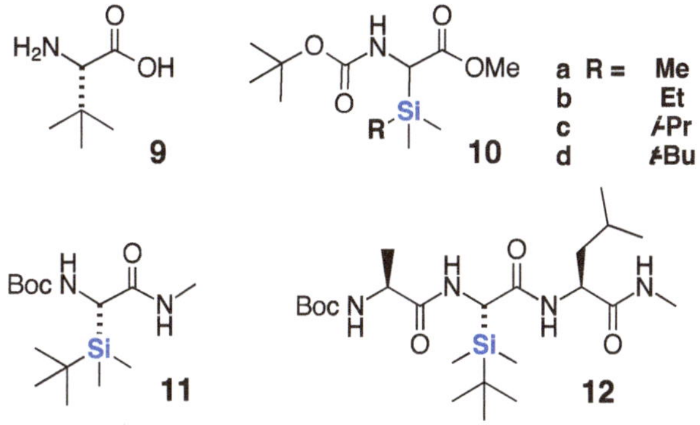

8.3 Silicon Containing Amino Acids [2]

8.3.1 The α-Silyl Amino Acids

Amino acids with an organosilicon substituent attached through silicon at the alpha position, e.g. compound **10** Fig. 8.3, have been the most difficult of the silicon substituted amino acids to prepare. Silicon analog **10a** of the unnatural amino acid *tert*-leucine **9** has not been isolated as the parent amino acid, but has been prepared with the amine protected and with the acid in the form of esters and amides.

The difficulty of working with amino acids **10** is exemplified by trimethylsilyl **10a**. This substance, when dissolved in methanol-d_4 at 20 °C, undergoes methanolysis of the Si–C bond with a half life of less than 12 h, yielding the protected glycine and trimethylsilyl methyl ether. Increasing the size of the silicon substituent leads to an increase in stability. Addition of a single CH_2 group, ester **10b**, increases the methanolysis half life nearly 20 fold, to 23 days, a trend that continues with **10c** and **10d**. Changing the ester **10d** to a primary amide **11** leads to another significant increase in stability. Polypeptides such as **12** have been prepared. The stability of the Si–C bond in these structures remains, however, problematic.

Two major approaches have been developed for the synthesis of these silane amino acid derivatives, Fig. 8.4 and Fig. 8.5. Reverse-aza-Brook rearrangement can be a useful method for migrating a silicon group from nitrogen to the adjacent carbon, Fig. 8.4. For compound **13**, the Boc group acts as a metallation directing group and the alkene is adequate to stablize the intermediate carbanion. When the metallation is conducted with (–)-sparteine as an additive, the product **14** is produced with 95 % ee. Oxidation then leads to acid **15**. Unfortunately, the alkene oxidation is accomplished in good yield only when the silicon carries *tert*-butyl substitution. Less robust silyl amino acid derivatives can be prepared using furfuryl amine derivative **16**. This transformation of **16** to **17** illustrates a single-flask operation for introducing the silane. When the furan **17** is subjected to ozonolysis a carboxylic acid is produced directly, and this process can be used to prepare even the sensitive trimethylsilyl substituted products such as **10a**.

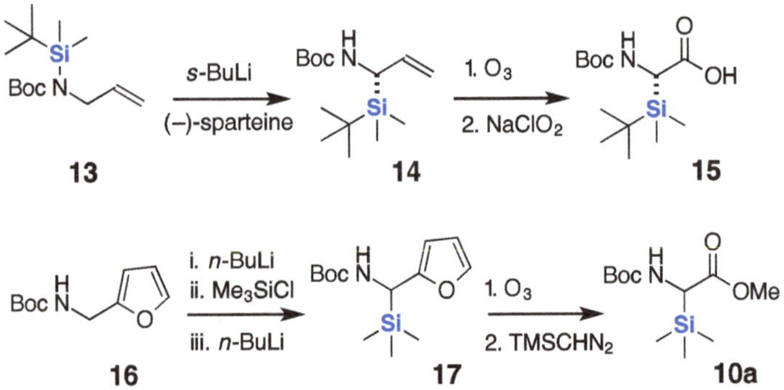

Fig. 8.4 α-Silyl amino acids by reverse-aza-Brook rearrangement

Fig. 8.5 α-Silyl amino acids by insertion into N–H bonds

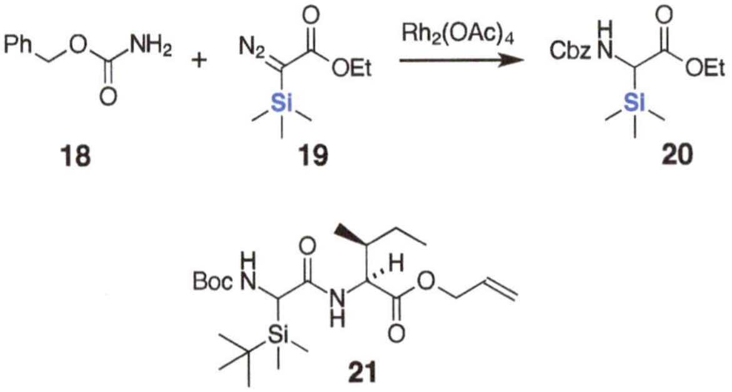

A versatile and efficient method for constructing these α-silyl amino acids involves rhodium catalyzed insertion of a carbenoid into N–H bonds such as with carbamate **18**, Fig. 8.5 [30]. A variety of *N*-protected α-silyl amino acids can be prepared this way. Dipeptide **21** was prepared using this technique.

Few additional reports have appeared concerning the preparation or use of these amino acids.

8.3.2 The β-Silyl Amino Acids

If silicon has two or more saturated carbons between it and a carbonyl group, such as the β-silyl amino acid **1**, the structure is far more stable than the α-silyl amino acids such as **10**. The original, racemic preparation of β-(trimethylsilyl)alanine **1** used classical malonate alkylation chemistry and the product withstands vigorous treatment with both acid and base (however, see Fig. 8.8) [1]. Asymmetric synthesis of **1** has been accomplished numerous times and three of the asymmetric methods are shown in Fig. 8.6. Each of these employs alkylation with a (halomethyl)trimethylsilane, followed by hydrolysis [31–33].

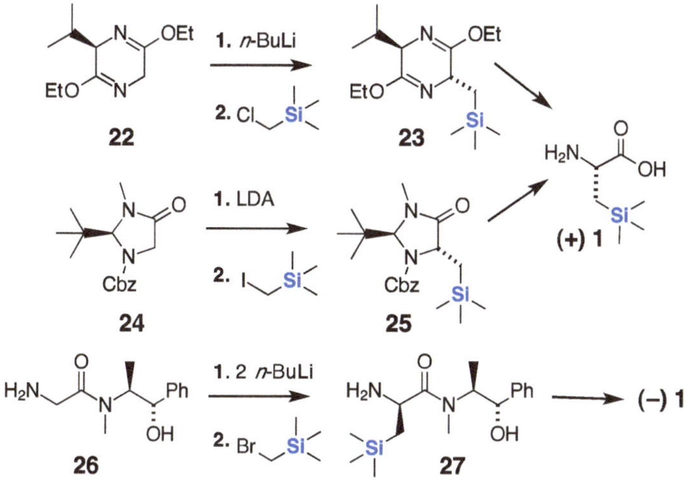

Fig. 8.6 Three methods for asymmetric synthesis of β-(trimethylsilyl)alanine

Fig. 8.7 Synthesis of 3-(dimethylsilyl)proline

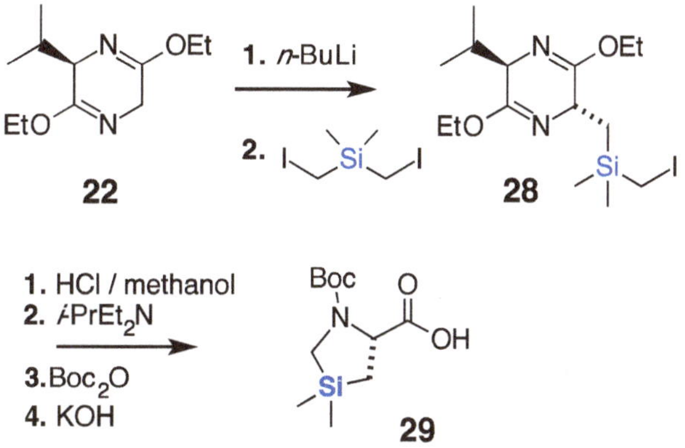

The Schöllkopf reagent **22** can also be used to prepare silaproline **2** in enantiopure form. Deprotonation and alkylation with di(iodomethyl)dimethylsilane gives **28**, Fig. 8.7. When this product is hydrolyzed and neutralized it cyclizes to form the silaproline structure. Protection of the amine and hydrolysis of the ester yields *N*-protected silaproline **29** [34].

The Si–C bonds in these β-silyl amino acids are robust, however if the groups attached to silicon are unsaturated they can suffer bond cleavage, Fig. 8.8. Subjecting racemic **30** to standard ester hydrolysis conditions (refluxing 6 N HCl for 1 h) led to cleavage of the silicon–phenyl bond. Isolation of the resulting product gave disiloxane **32** as a mixture of diastereomers. When this amino acid dimer was dissolved in D_2O only the monomeric **31** was detected, demonstrating reversibility of the siloxane formation [35]. For an additional example of a silanol amino acid, see Fig. 8.13.

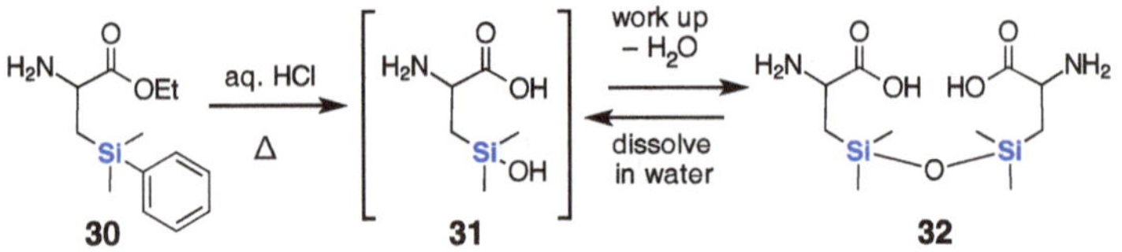

Fig. 8.8 Preparation of 3-(hydroxydimethylsilyl)alanine

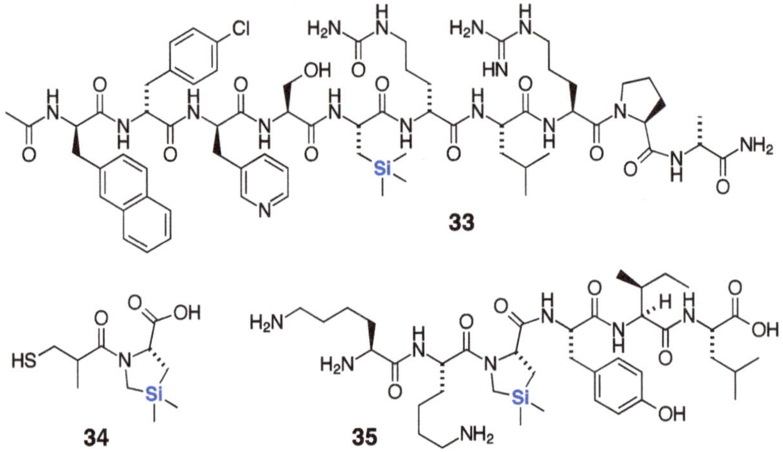

Fig. 8.9 Bioactive peptides incorporating silane amino acids

Incorporation of these β-silyl amino acids into bioactive structures has produced useful properties. Replacement of a tyrosine in the decapeptide Cetrorelix with β-(trimethylsilyl)alanine gave **33** [31]. Cetrorelix is an antagonist of gonadotropin-releasing hormone useful in cancer therapy [36]. Receptor binding assays using **33** and its carbon analog (containing a *tert*-butyl in place of the trimethylsilyl group) found them to be potent antagonists. When tested *in vivo* with rats, the silane **33** was found to lower both testosterone and luteinizing hormone levels longer than the carbon analog [31].

Substitution of proline in the angiotensin-converting enzyme (ACE) inhibitor Captopril with 3-(dimethylsilyl)proline **2** gave **34**, an inhibitor of ACE nearly as potent as Captopril itself [37]. Inhibition of ACE is an important treatment for hypertension.

Insertion of 3-(dimethylsilyl)proline into into a neurotensin analog produced **35**. This structure retained biological activity and was much more stable toward enzymatic degradation than nonsilane congeners [3].

A variety of amino acids more remotely substituted with silicon have been prepared and reviews of that work can be found elsewhere [2].

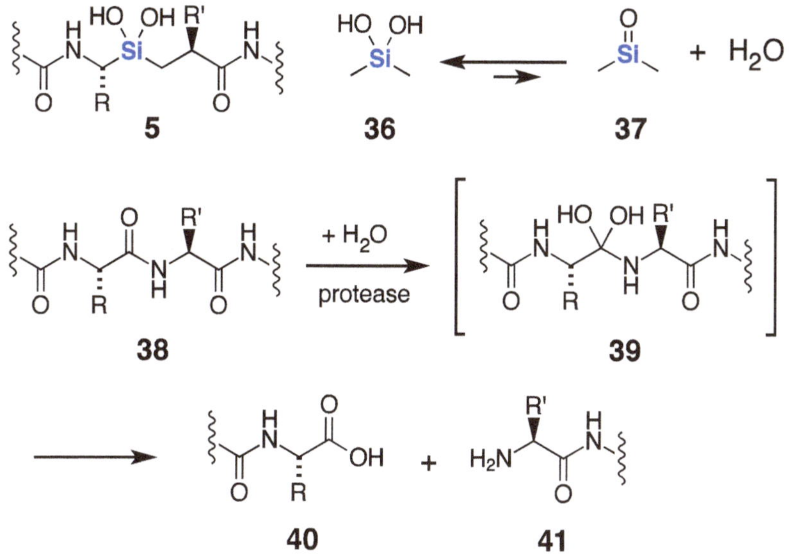

Fig. 8.10 Protease enzymes mediate many biological processes by specific peptide cleavages. Silanediol **5** mimics bound intermediate **39** and can inhibit the enzymes

8.4 Silanediol Protease Inhibitors

8.4.1 Design and Activity

Silicon can also be part of a peptide backbone, e.g. **5**, Fig. 8.10. Silanediols such as **5** are analogs of a hydrated carbonyl. Silanediols like **36** are stable "hydrates" because double bonds to silicon (e.g., **37**) are formed only under extreme or unusual circumstances [38]. Within a peptide-like structure such as **5**, the silanediol can masquerade as hydrated amide **39**, an intermediate structure in the hydrolysis of peptides. Hydrated amide analogs can bind to a protease active site and act as inhibitors [39]. Proteases are a large class of enzymes that mediate a broad spectrum of biological processes by cleavage of peptide bonds. Important pharmaceutical applications of protease inhibitors include the treatment of hypertension (by inhibition of angiotensin-converting enzyme [40] and renin [41]), cancer (inhibition of the proteosome [42]) and AIDS (inhibition of the HIV protease [43]) to name a few [44].

Whereas silanediols are stable toward formation of silanones **37**, their best known attribute is an alternative dehydration by polymerization, forming siloxanes (silicones) such as **42**, Fig. 8.11. In the case of dimethylsilanediol **36**, this polymerization is spontaneous and the product siloxanes are prized for their stability [45]. This polymerization, however, is sensitive to the steric environment of the silane. An outstanding example of this steric effect is diisobutylsilanediol, a liquid crystalline material with little tendency to oligomerize [46].

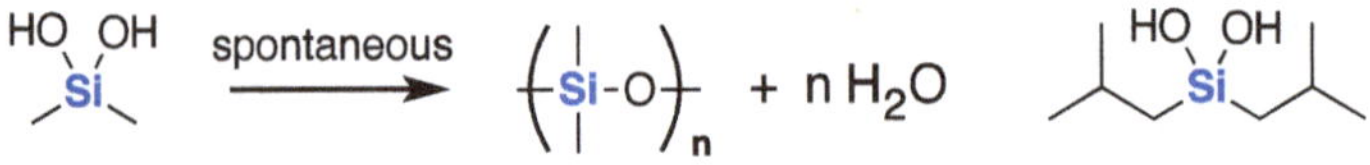

Fig. 8.11 Stability of silanols toward self condensation is structure dependent

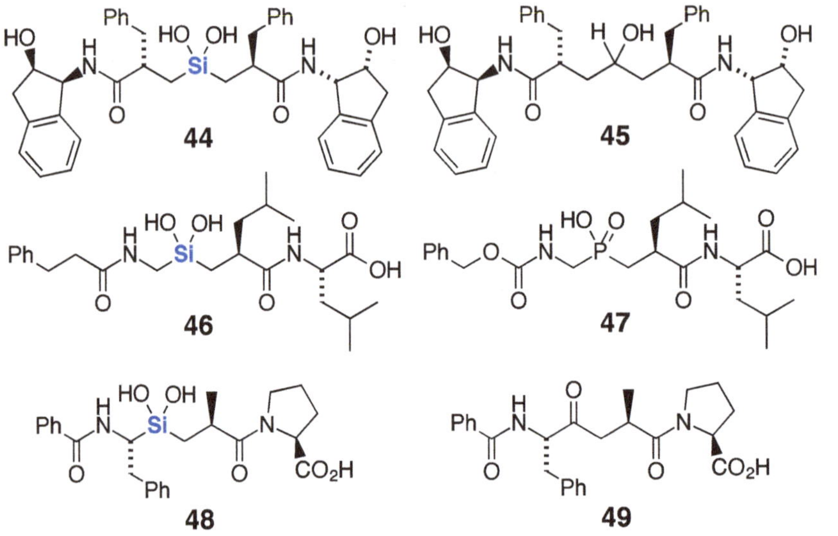

Fig. 8.12 Silanediol protease inhibitors

Three silanediol protease inhibitors are shown in Fig. 8.12, in order of increasing complexity: the C_2 symmetric **44** has identical organic substituents on the silandiol, silanediol **46** has two different organic structures on the silicon and **48** has very different structures on each side of the silanediol group, each with stereochemistry.

The symmetry of **44** matches the (unusual!) C_2 symmetric HIV protease, and was modeled after the corresponding carbinol **45** reported by Merck [47]. Testing of **44**, **45** and the commercial HIV protease inhibitor indinavir (not shown) found them to inhibit the enzyme to a similar degree, with K_i values of 2.7 nM, 0.38 nM and 0.37 nM, respectively [48]. In addition, IC_{90} values for protection of whole cells against HIV infection gave parallel values, indicating that they all penetrated cell membranes to the same degree, thereby demonstrating the drug-like properties of silanediol **44**.

Phosphinic acid **47** was the starting point for design of silanediol **46**, two structures that inhibit the metalloprotease thermolysin [49]. Compounds **46** and **47** have thermolysin inhibition K_i values of 41 nM and 10 nM, respectively. The Cbz group on the amine of **47** was replaced by a dihydrocinnamoyl group in **46** after the discovery that a Cbz group did not withstand the strongly acidic conditions used to deprotect the silanediol precursor (see below) [50]. A crystal structure of **46** bound to the active site of thermolysin was determined by Juers and Matthews and showed the anticipated interaction of the silanediol with the active site zinc ion [51].

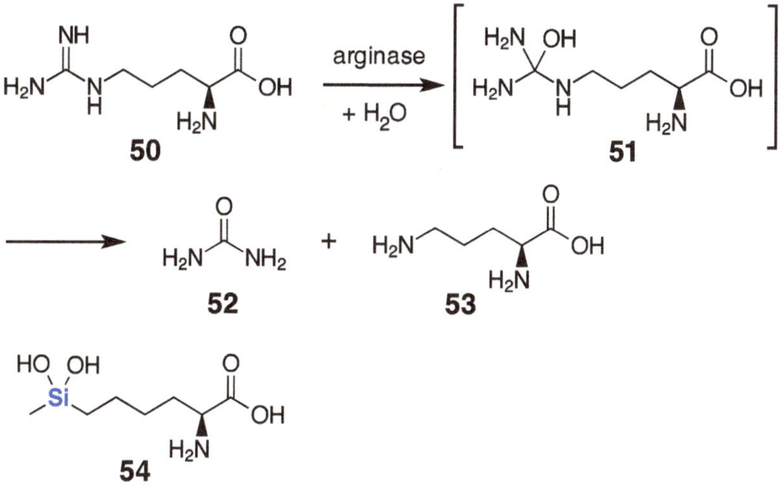

Fig. 8.13 Sterically unhindered, water soluble **54**, was an ineffective inhibitor of arginase

The most complex of the silanediols, in terms of the core silanediol amino acid component, is **48**, an inhibitor of the enzyme angiotensin-converting enzyme (ACE). Inhibitor **48** was modeled after ketone inhibitor **49**, with K_i values of 3.8 nM and 1.0 nM, respectively [52]. Four diastereomers of both **48** and **49** were prepared, changing the stereogenic centers flanking the silane and the ketone, and their potencies as ACE inhibitors were evaluated [53].

These three silanes, all successful as protease inhibitors, suggest that incorporation of the silanediol group can be a useful drug design strategy.

The least sterically hindered silanediol amino acid is **54**, which was proposed as an inhibitor of arginase, Fig. 8.13. This enzyme catalyzes the hydration and hydrolysis of the guanidine unit of arginine **50**, producing urea **52** and ornithine **53**. Arginase is a pharmaceutical target because of its role in regulating NO production [54, 55]. Silanediol **54** was, however, not effective as a mimic of hydrated guanidine **51** and did not inhibit the enzyme. Nevertheless, silane **54** is of interest for its relationship to silanol **31** (Fig. 8.8) and as a tool for understanding the role of water solubility in modulating siloxane formation [56].

8.4.2 Silanediol Synthesis

To assemble the silanediols **5**, a protecting group was needed, Fig. 8.14. The ability of unsaturated groups on silicon to undergo "protodesilylation" under acidic conditions, fundamentally a hydrolysis reaction, seemed a reasonable approach and phenyl was selected as a suitably robust group to carry through synthetic sequences [57, 58]. Strongly acidic conditions (e.g., triflic acid in trifluoroacetic acid) has been widely used for deprotection of synthetic peptides [59]. Protodesilylation is

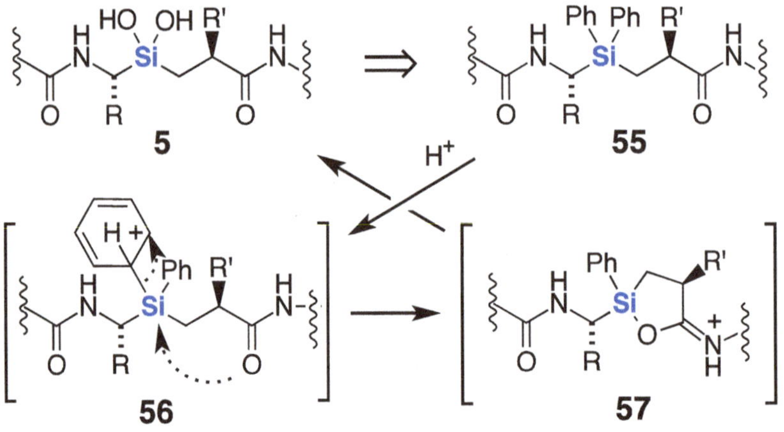

Fig. 8.14 Phenyl is an acid-labile protecting group for the silanediols

also a classic electrophilic aromatic substitution reaction, promoted by the ability of silicon to stabilize a β-cation (**56**), and the reactivity of the aromatic ring can be altered by ring substitution [60, 61]. Use of diphenylsilyl as a silanediol precursor turned out to be a fortuitous choice because of the importance of phenyl groups in the formation and stabilization of silane anions (see Fig. 8.23) [62, 63]. Studies suggest that the amide groups flanking the diphenylsilane participate in the loss of phenyl, presumably by attacking the protonated arylsilane **56**, Fig. 8.14. In cases without amides on both sides of the silane, loss of the second phenyl group is slower and more difficult [64]. Triflic acid has been routinely used for this deprotection step, but recently methods employing the much milder trifluoroacetic acid have been enumerated [61].

8.4.3 An HIV Protease Inhibitor

HIV protease inhibitor **44** (Fig. 8.12) was prepared in enantiomerically pure form by addition of two equivalents of an optically pure lithium reagent to a dichlorosilane. Starting with the Evans chiral auxiliary **58** and alkylation of the titanium enolate with BOMCl, Fig. 15, reductive cleavage of the auxiliary and conversion of the resulting alcohol to an iodide gave **59** (76 % for 3 steps). Metal–halogen exchange followed by addition of 0.45 equivalents of diphenyldichlorosilane gave **60** in 99 % yield. Cleavage of the benzyl ethers and a two-stage oxidation of the resulting diol gave the diacid (84 %). Coupling of this diacid with commercially available amino alcohol **62** using diethylphosphoryl cyanide gave diamide **61** (75 %). The critical deprotection of diphenylsilane **61** to give the diol **44** was carefully studied with regard to cleavage of both silicon–phenyl bonds and isolation of monomer **44**. Pure **44** was most easily isolated by precipitation, in 37 % yield [48].

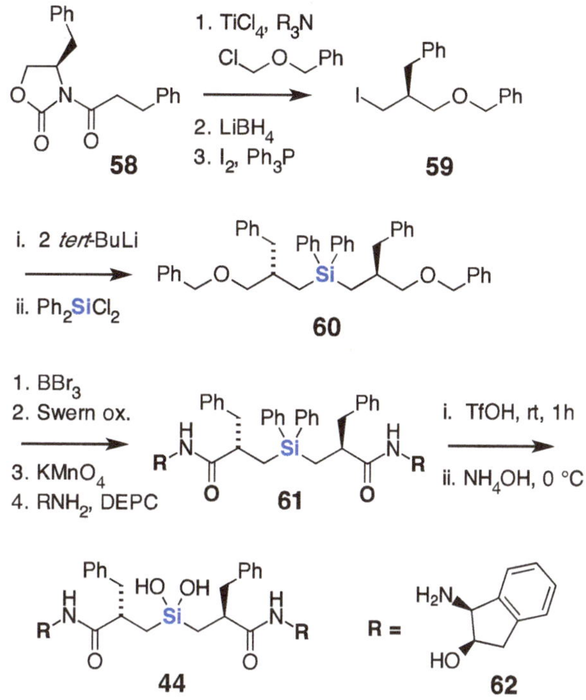

Fig. 8.15 Synthesis of silanediol HIV inhibitor **44**

8.4.4 Thermolysin Inhibitor

Preparation of a silanediol inhibitor for thermolysin, structure **46**, with two different organic groups attached to silicon, started with a chloromethyl silane and used an optically active organolithium reagent, similar to the procedure developed for HIV inhibitor **44** (Fig. 8.15). Starting with commercially available chloromethyl trichlorosilane **63**, treatment with phenylmagnesium chloride followed by a 48 % HF work up led to fluorosilane **64**, Fig. 8.16. As described by Eaborn, despite the very high strength of the Si–F bond, fluorosilanes react readily with nucleophiles [65]. Moreover, mono- and difluorosilanes are stable to moisture, making them much easier to handle than chlorosilanes. (note that fluorosilanes have the potential to generate HF burns and proper precautions must be taken!)

Treatment of silane **64** with enantiomerically pure lithium reagent **65** gave chloromethylsilane **66**. Displacement of the chloride by phthalimide, removal of the benzyl ether and oxidation led to acid **67**. Coupling with leucine *tert*-butyl ester, removal of the phthalimide group and coupling with dihydrocinnamoyl chloride gave penultimate product **68**.

Silanediol **46** was the least sterically hindered protease inhibitor prepared and therefore had the most potential to polymerize. Subjection of **68** to the procedure previously used for conversion of the diphenylsilyl to a dihydroxysilyl group, triflic acid followed by ammonium hydroxide neutralization, led to substantial amounts

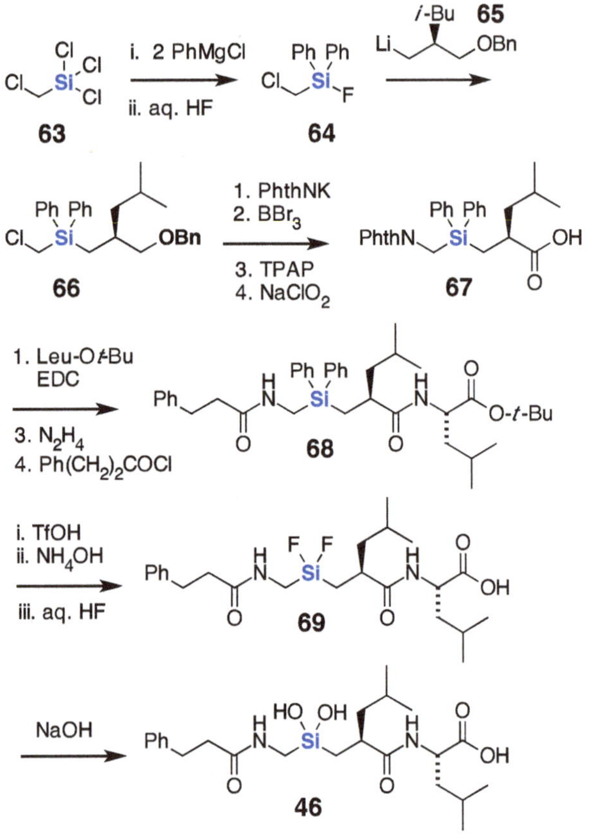

Fig. 8.16 Synthesis less hindered silanediol 46 led to the use of a difluorosilane intermediate **69**

of siloxane oligomers. It was this oligomerization difficulty that led to the development of a useful variation in the hydrolysis procedure: following cleavage of the aryl groups from silicon with triflic acid and ammonium hydroxide to hydrolyze the (presumed) cyclized intermediate, the resulting crude mixture was then diluted with 48 % HF. As anticipated, aqueous HF converted all of the silicon–heteroatom bonds, including siloxanes, into the crystalline difluorosilane monomer **69**. Suspending **69** in water and addition of sodium hydroxide led to a rapid dissolution and clean formation of a water soluble, monomeric silanediol [50].

8.4.5 Angiotension-Converting Enzyme Inhibitors

Silanediol inhibitors of angiotensin-converting enzyme (ACE) were the most complex synthesis challenges among the silanediols, and were also the first silanediol inhibitors targeted. The first of the silanediol inhibitors, **70** and **71**, Fig. 8.17, were prepared with no control of stereochemistry [66]. Structure **70** was a mixture of

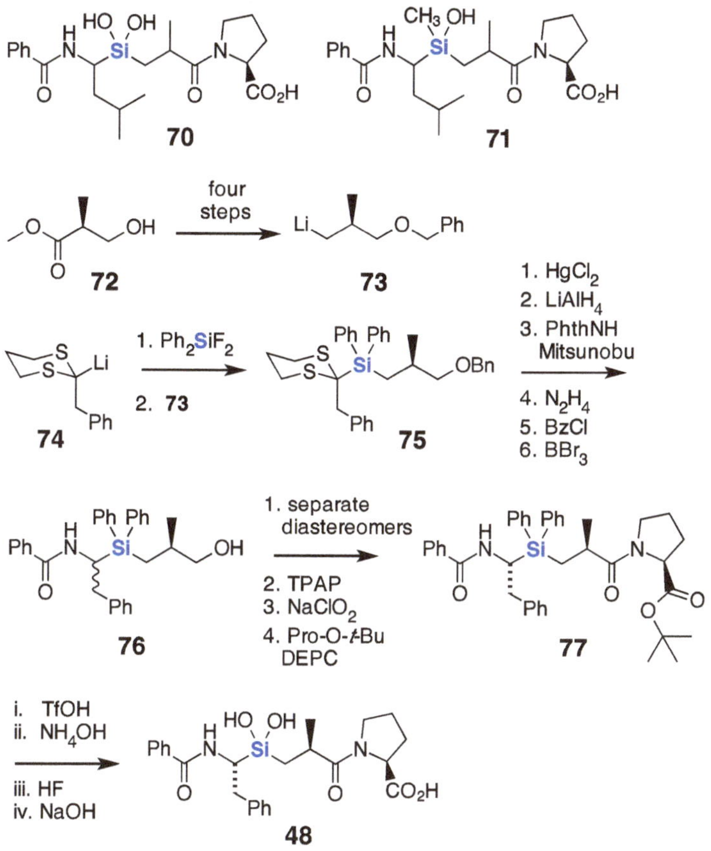

Fig. 8.17 Synthesis of ACE inhibitor **48** with two stereogenic centers flanking the silane

four stereoisomers, whereas **71** also has a stereogenic silicon and was a mixture of eight stereoisomers. The extremely poor inhibition shown by methylsilanol **71** (IC_{50}>3000 nM) relative to **70** (IC_{50} 14 nM) was evidence that a silanediol group was important for enzyme inhibition by these structures [52].

All four stereoisomers of **48** were prepared, keeping D-proline constant, and the synthesis of one of these is outlined in Fig. 8.17. The stereochemistry of the methyl substituent in **48** was obtained from the (*S*)-Roche ester **72**, converting it to optically active lithium reagent **73**. Coupling difluorodiphenylsilane with metallated dithiane **74** followed by displacement of the second fluoride by **73** gave **75** which contains all the carbons of silane **48**. Conversion of the dithiane to the amine involved hydrolysis and reduction of the ketone, without control of stereochemistry. In this case, both stereoisomers were needed, but it was recognized that in future work stereocontrol of this transformation needed to be addressed (see Fig. 8.21).

Mitsunobu displacement of the alcohol with phthalimide, removal of the phthalic acid group and coupling of the resulting primary amine with benzoyl chloride completed that portion of the molecule. Removal of the benzyl group then gave a mixture of two diastereomers **76** that could be separated by column chromatography. Each was taken on separately. Oxidation of the alcohol, condensation with proline *tert*-butyl ester and then “standard” hydrolysis of the diphenylsilane to the silanediol completed the sequence.

The chemistries described in Figs. 8.15, 8.16, 8.17 successfully assembled the protease inhibitors, yet the length of the syntheses left much room for improvement. A more efficient and general set of methods for assembling these structures was highly desirable and these efforts are discussed below.

8.5 Improved Chemistries for Silanediol Inhibitor Construction

Embedded in the silanediol protease structure **5** are two challenges, the α-amino silane **78** and the β-silyl propionic acid **79**, both with stereogenic centers, Fig. 8.18. Methods for efficient assembly of each of these substructures have been investigated, and are discussed separately, below.

8.5.1 α-Alkyl-β-Silyl Acids

In Figs. 8.15, 8.16, 8.17, the α-alkyl-β-silyl fragment of the inhibitors (**79**) was either derived from the chiral pool (**72**, Fig. 8.17) [67] or through use of Evans chiral auxiliary technology [68]. A number of improvements have been devised.

8.5.1.1 Asymmetric Hydroboration of 2,5-Dihydrosiloles

The heterocycle 2,5-dihydro-3-methyl-1,1-diphenylsilole **80** and related structures are readily prepared in large quantities (>300 g [69]), in this case from dichlorodiphenylsilane and isoprene, Fig. 8.19. This distillable product was found to undergo asymmetric hydroboration with Brown’s isopinocamphylborane to give alcohol **81** in high ee. This alcohol, when treated with aqueous HF in ethanol at reflux, undergoes a near quantitative Peterson fragmentation to give fluorosilane **83** with stereochemistry of the methyl group set [70]. The fluoride in **83** undergoes high yield displacement with a broad range of nucleophiles, including sensitive chloromethyllithium to give **84** Oxidative cleavage of the alkene forms the desired α-methyl-β-silyl propionic acid (see Fig. 8.19).

Isoprene is one of the few 2-alkyl-1,3-butadienes that is commercially available. To supplement this dihydrosilole work, a new method for preparation of 2-alkyl-1,3-dienes was developed, starting with commercially available dibromobutene **85** (also easily prepared by bromination of 1,3-butadiene) [71]. Copper (I)-mediated

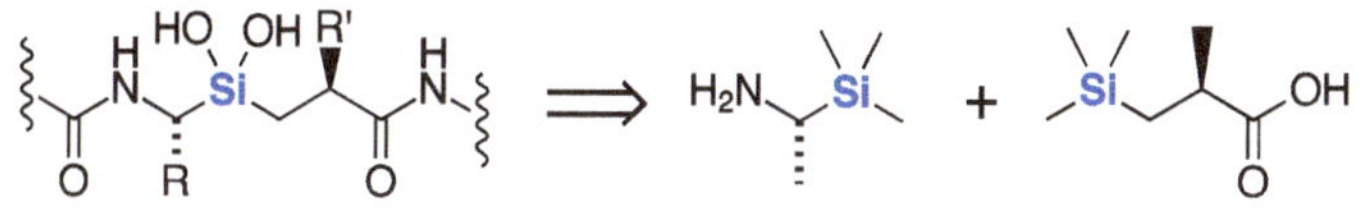

Fig. 8.18 The two challenges of silanediol protease inhibitor synthesis

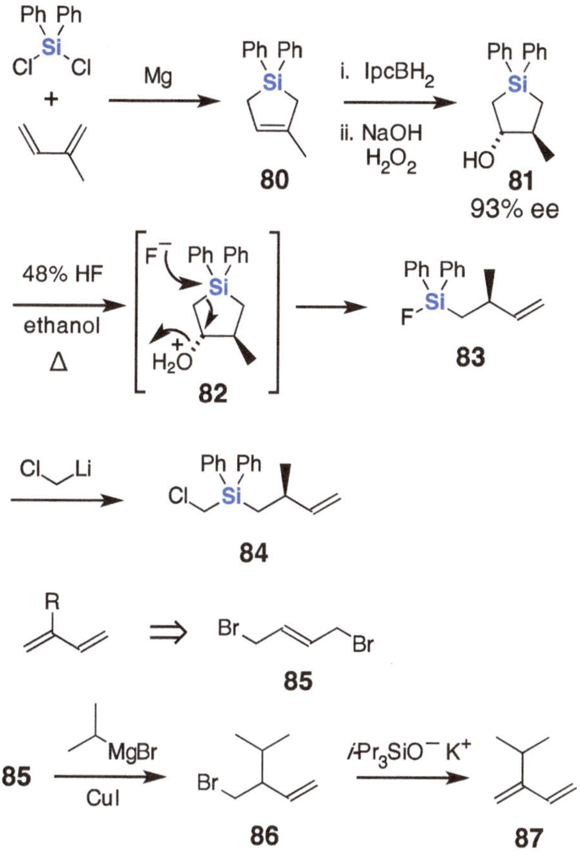

Fig. 8.19 Asymmetric hydroboration and cleavage of dihydrosiloles

S_N2' displacement of an allylic bromide yields **86** and dehydrohalogenation of **86** gives the 1,3-diene product **87**. The optimal condition for this elimination involved the use of Soderquist's silanolate reagent [72].

8.5.1.2 Asymmetric Intramolecular Hydrosilylation

One of the most powerful ways to make silicon–carbon bonds is hydrosilylation [73, 74]. Combining the two commercially available reagents chlorodiphenylsilane

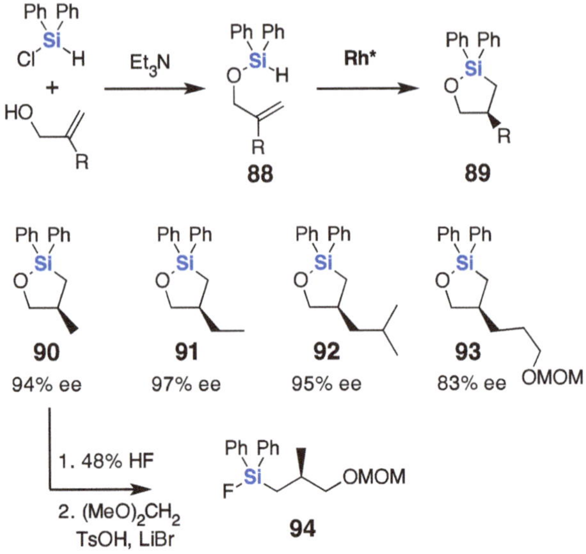

Ph Ph Si Cl H + HO R → Et$_3$N → Ph Ph Si O H R **88** → Rh* → Ph Ph Si O R **89**

90 94% ee; **91** 97% ee; **92** 95% ee; **93** 83% ee OMOM

90 → 1. 48% HF 2. $(MeO)_2CH_2$ TsOH, LiBr → Ph Ph Si F OMOM **94**

Fig. 8.20 Asymmetric intramolecular hydrosilylation sets stereochemistry with high ee

and methallyl alcohol (R = methyl), Fig. 8.18, gives silyl ether **79**. Treatment of this with a variety of metals will induce cyclization and formation of silafuran **89** [75, 76]. When rhodium is used with (*S, S*)-ethylferrotane [77] as a ligand, this cyclization sets the stereogenic center in **89** with good to excellent enantioselectivity [78].

One path to convert these silafurans to protease inhibitor structures involves opening of the ring with aqueous HF and protecting the resulting alcohol to give **94** [78]. These fluorosilanes react readily with nucleophiles and can also be converted into nucleophilic silyllithium reagents [70]. An even more efficient conversion of **94** to protease inhibitor structures is described in Fig. 8.23.

8.5.2 *α-Alkyl-α-Amino Silanes*

8.5.2.1 Asymmetric Reduction of a Silyl Ketone

Construction of the α-amino silanes component of the silanediol **46** with the correct stereochemistry was initially done by separation of diastereomers, following reduction of a silyl ketone without stereocontrol and displacement of the alcohol with phthalimide, Fig. 8.17. A better solution was to control the ketone reduction, and this has been done in the case of **95** with stoichiometric use of the CBS-borane complex, Fig. 8.21. Mitsunobu inversion then gave the desired stereochemistry (**97**). Oxidative cleavage of the alkene then completes the sequence [79].

Fig. 8.21 Control of stereochemistry by asymmetric reduction of the ketone

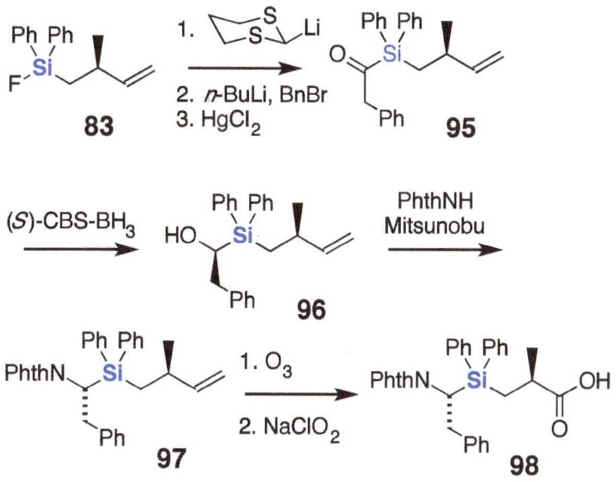

8.5.2.2 Asymmetric Reverse-aza-Brook Rearrangment

Another asymmetric method for assembling α-amino silanes uses metallation chemistry, Fig. 8.22. Silyl triflates react with carbamates to give *N*-silyl derivatives **100**. The Boc group in **100** then acts as a metallation-directing group. Treatment with *sec*-butyllithium leads to carbanion **101**, which undergoes rearrangement at low temperature to give **102** and **103** on workup. An anion-stabilizing group is required here; replacement of benzyl with ethyl gives no metallation but a vinyl group is sufficient to allow deprotonation to take place efficiently (e.g., **104**).

When the metallation step is conducted in the presence of (–)-sparteine, the product **105** is formed in > 90 % ee. The stereochemistry of the product using the natural isomer of sparteine is correct for α-silyl amino acid synthesis (see Fig. 8.4) but wrong for analogs of the naturally occurring peptides. Alternatives to sparteine that provide the other enantioselectivity during metallation reactions are known [80].

8.5.2.3 Silyllithium Addition to Sulfinimines

An efficient method for assembling the α-amino silane structure is addition of silyllithium reagents to sulfinimines. Skyrdstrup and Nielson were the first to demonstrate that this could be done in high yield, converting Ellman [81] sulfinimine **106** to **108** in 76 % yield, with diastereoselectivity > 95 % [64, 82].

The subsequent discovery that silyl ether **90** could be converted to dianion **109** by treatment with lithium, lent another level of efficiency to the overall procedure [78]. Silane **90** is formed with high enantioselectivity by hydrosilylation (Fig. 8.20). Addition of dianion **109** to Davis [83] sulfinimine **110** gave alcohol **111** in 76 % yield. The combination of asymmetric hydrosilylation to set the stereochemistry of **90** and the use of the sulfinimine method to set the amino group stereochemistry makes for a highly efficient and versatile procedure.

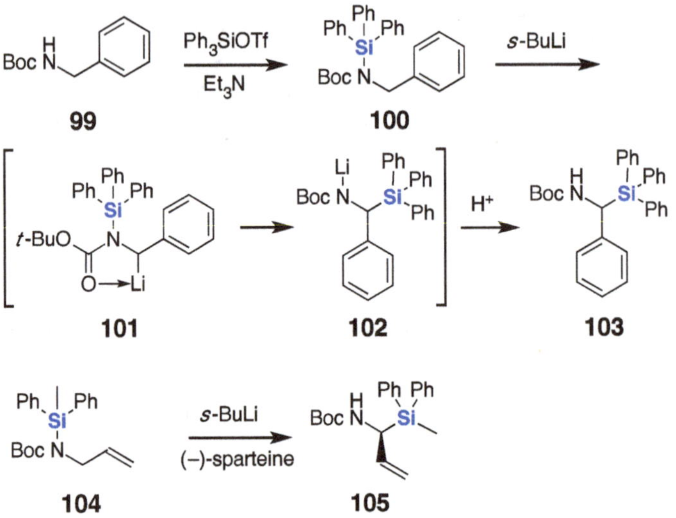

Fig. 8.22 Reverse-aza-Brook rearrangement of *N*-silyl carbamates

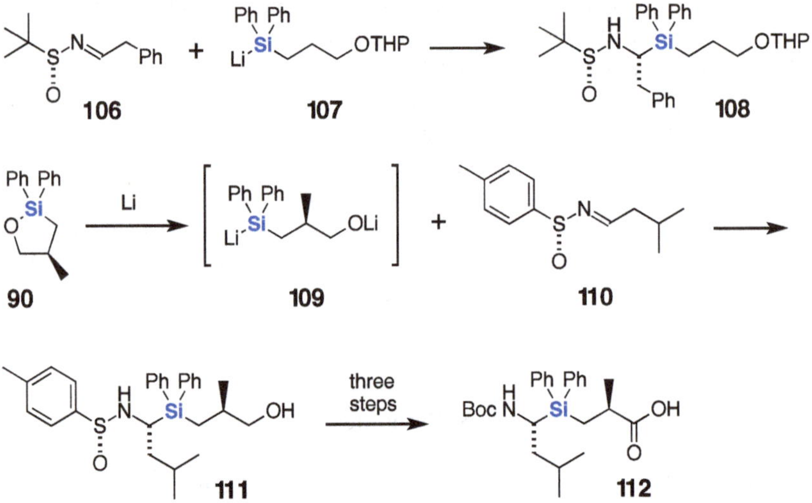

Fig. 8.23 Sulfinimines undergo addition of silyl anions with high diastereoselectivity

8.6 Future Prospects

Silicon, with its close relationship to carbon, will continue to be a part of the innovation in bio-active substance design and development. As the pharmaceutical industry moves more and more into peptide and protein based therapies, silicon-containing amino acids and peptides will be part of those investigations. Use of the silicon amino acids holds great promise for maintaining biological activity while simultaneously suppressing the rapid degradation of peptides *in vivo* (see Fig. 8.9). The growing commercial availability of silicon containing amino acids will help fuel that process.

The more recent silanediol peptidomimetics, which act as inhibitors of hydrolase enzymes, are expected to see much more development as their properties and applications continue to be explored.

References

1. Birkofer L, Ritter A (1956) Angew Chem 68:461–462
2. Mortensen M, Husmann R, Veri E, Bolm C (2009) Chem Soc Rev 38:1002–1010
3. Cavelier F, Vivet B, Martinez J, Aubry A, Didierjean C, Vicherat A, Marraud M (2002) J Am Chem Soc 124:2917–2923
4. Liu G, Sieburth SMcN (2005) Org Lett 7:665–668
5. Sun H, Martin C, Kesselring D, Keller R, Moeller KD (2006) J Am Chem Soc 128:13761–13771
6. Sun H, Moeller KD (2003) Org Lett 5:3189–3192
7. Sun H, Moeller KD (2002) Org Lett 4:1547–1550
8. Fessenden RJ, Fessenden JS (1967) Adv Drug Res 4:95–132
9. Tacke R, Wannagat U (1979) Top Curr Chem 84:1–75
10. Fessenden RJ, Fessenden JS (1980) Adv Organomet Chem 18:275–299
11. Barcza S (1988) In: Corey ER, Corey JY, Gaspar P (eds) The value and new directions of silicon chemistry for obtaining bioactive compounds in silicon chemistry. Wiley, New York, pp 135–144
12. Tacke R, Linoh H (1989) In: Patai S, Rappoport Z (eds) Bioorganosilicon chemistry in the chemistry of organic silicon compounds. Wiley, New York, pp 1143–1206
13. Bains W, Tacke R (2003) Curr Opin Drug Discov Dev 6:526–543
14. Mills JS, Showell GA (2004) Expert Opin Investig Drugs 13:1149–1157
15. Englebienne P, Hoonacker AV, Herst CV (2005) Drug Des Rev Online 10:654–671
16. Sieburth SMcN (2008) Bioactive organosilanes in gelest catalog. pp 76–83.
17. Lazareva LF (2011) Russ Chem Bull Int Ed 60:615–632
18. Fessenden RJ, Hartman RA (1970) J Med Chem 13:52–54
19. Fleming I, Dunoguès J, Smithers R (2004) The electrophilic substitution of allylsilanes and vinylsilanes in organic reactions. Wiley, New York (1989), 37:57–575
20. Franz AK, Woerpel KA (2000) Acc Chem Res 33:813–820
21. Denmark SE, Wehrli D, Choi JY (2000) Org Lett 2:2491–2494
22. Sunderhaus JD, Lam H, Dudley GB (2003) Org Lett 5:4571–4573
23. Ottosson H, Steel PG (2006) Chemistry 12:1576–1585
24. Brändström A (1959) Acat Chem Scand 13:619–622
25. Woo DV, Christian JE, Schnell RC (1979) Can J Pharm Sci 14:12–14
26. Ohshita J, Tokunaga Y, Sakurai H, Kunai A (1999) J Am Chem Soc 121:6080–6081
27. Tacke R, Popp F, Müller B, Theis B, Burschka C, Hamacher A, Kassack MU, Schepmann D, Wünsch B, Jurva U, Wellner E (2008) Chem Med Chem 3:152–164
28. Tacke R, Heinrich T, Bertermann R, Burschka C, Hamacher A, Kassack MU (2004) Organometallics 23:4468–4477
29. Johansson T, Weidolf L, Popp F, Tacke R, Jurva U (2010) Drug Metab Disp 38:73–83
30. Bolm C, Kasyan A, Drauz K, Günther K, Raabe G (2000) Angew Chem Int Ed 39:2288–2290

31. Tacke R, Merget M, Bertermann R, Bernd M, Beckers T, Reissmann T (2000) Organometallics 19:3486–3497
32. Fitzi R, Seebach D (1988) Tetrahedron 44:5277–5292
33. Myers AG, Gleason JL, Yoon T, Kung DW (1997) J Am Chem Soc 119:656–673
34. Vivet B, Cavelier F, Martinez J (2000) Eur J Org Chem 5:807–811
35. Tacke R, Schmid T, Merget M (2005) Organometallics 24:1780–1783
36. Reissmann T, Schally AV, Bouchard P, Riethmüller H, Engel J (2000) Hum Reprod Update 6:322–331
37. Dalkas GA, Marchand D, Galleyrand J-C, Martinez J, Spyroulias GA, Cordopatis P, Cavelier FJ (2010) Peptide Sci 16:91–97
38. Tokitoh N, Okazaki R (1998) In: Apeloig Y, Rappoport Z (eds) Recent advances in the chemistry of silicon-heteroatom multiple bonds in the chemistry of organic silicon compounds II. Wiley, New York, pp 1063–1103
39. Babine RE, Bender SL (1997) Chem Rev 97:1359–1472
40. Matchar DB, McCrory DC, Orlando LA, Patel MR, Patel UD, Patwardhan MB, Powers B, Samsa GP, Gray RN (2008) Ann Intern Med 148:16–29
41. Gradman AH, Pinto R, Kad R (2008) Curr Opin Pharmacol 8:120–126
42. Groll M, Berkers CR, Ploegh HL, Ovaa H (2006) Structure 14:451–456
43. Erickson JW, Eissenstat MA (1999) In: Dunn BM (ed) HIV protease as a target for the design of antiviral agents for AIDS in proteases of infectious agents. Academic Press, New York, pp 1–60
44. Dunn B (2012) Proteinases as drug targets. The Royal Society of Chemistry, London
45. Lebrun JJ, Porte H (1991) Polysiloxanes. In: Bevington JC, Allen G (eds) Comprehensive polymer science V. Pergamon, New York, pp 593–609
46. Bunning JC, Lydon JE, Eaborn C, Jackson PM, Goodby JW, Gray GW (1982) J Chem Soc Faraday Trans 1(78):713–724
47. Bone R, Vacca JP, Anderson PS, Holloway MK (1991) J Am Chem Soc 113:9382–9384
48. Chen C-A, Sieburth SMcN, Glekas A, Hewitt GW, Trainor GL, Erickson-Viitanen S, Garber SS, Cordova B, Jeffry S, Klabe RM (2001) Chem Biol 8:1161–1166
49. Kim J, Glekas A, Sieburth SMcN (2002) Bioorg Med Chem Lett 12:3625–3627
50. Kim J, Sieburth SMcN (2004) J Org Chem 69:3008–3014
51. Juers D H, Kim J, Matthews B W, Sieburth SMcN (2005) Biochemistry 44:16524–16528
52. wa Mutahi M, Nittoli T, Guo L, Sieburth SMcN (2002) J Am Chem Soc 124:7363–7375
53. Kim J, Hewitt G, Carroll P, Sieburth SMcN (2005) J Org Chem 70:5781–5789
54. Christianson D W (2005) Acc Chem Res 38:191–201
55. Decaluwé K, Pauwels B, Verpoest S, Van de Voorde J (2011) J Sex Med 8:3271–3290
56. Kim JK, Sieburth SMcN (2012) J Org Chem 77:7701–7706
57. Fleming I, Henning R, Plaut H (1984) Chem Commun 29–31
58. Uhlig W (1996) Chem Ber 129:733–739
59. Atherton E, Sheppard R C (1989) Solid phase peptide synthesis: a practical approach. IRL, Oxford, pp 13–23
60. Eaborn C J (1956) J Chem Soc 4858–4864
61. Hernaỉndez D, Mose R, Skrydstrup T (2011) Org Lett 13:732–735
62. Lickiss PD, Smith CM (1995) Coord Chem Rev 145:75–124
63. Sekiguchi A, Lee VY, Nanjo M (2000) Coord Chem Rev 210:11–45
64. Nielsen L, Lindsay KB, Faber J, Nielsen NC, Skrydstrup TJ (2007) Org Chem 72:10035–10044
65. Eaborn CJ (1952) J Chem Soc 2840–2846
66. Sieburth SMcN, Nittoli T, Mutahi AM, Guo L (1998) Angew Chem Int Ed 37:812–814
67. Blaser H U (1992) Chem Rev 92:935–952
68. Evans DA, Urpi F, Somers TC, Clark JS, Bilodeau MT (1990) J Am Chem Soc 112:8215–8216
69. Mignani S, Damour D, Bastart J, Manuel G (1995) Synth Commun 25:3855–3861
70. Sen S, Purushotham M, Qi Y, Sieburth SMcN (2007) Org Lett 9:4963–4965

71. Sen S, Singh S, Sieburth SMcN (2009) J Org Chem 74:2884–2886
72. Soderquist JA, Vaquer J, Diaz MJ, Rane AM, Bordwell FG, Zhang S (1996) Tetrahedron Lett 37:2561–2564
73. Roy A K (2007) Adv Organomet Chem 55:1–59
74. Ojima I, Li Z, Zhu J (1998) Recent advances in the hydrosilylation and related reactions, In: Apeloig Y, Rappoport Z (eds) The chemistry of organic silicon compounds II. Wiley, New York, pp 1687–1792
75. Mironov VF, Kozlikov VL, Fedotov NS (1969) Zh Obshch Khim 39:966–970
76. Bergens SH, Noheda P, Whelan J, Bosnich B (1992) J Am Chem Soc 114:2128–2135
77. Berens U, Burk MJ, Gerlach A, Hems W (2000) Angew Chem Int Ed 39:1981–1984
78. Bo Y, Singh S, Duong H Q, Cao C, Sieburth SMcN (2011) Org Lett 13:1787–1789
79. Qi Y, Singh S Research Notes. Temple University
80. Dearden MJ, Firkin CR, Hermet JR, O'Brien PJ (2002) Am Chem Soc 124:11870–11871
81. Robak MT, Herbage MA, Ellman JA (2010) Chem Rev 110:3600–3740
82. Nielsen L, Skrydstrup T (2008) J Am Chem Soc 130:13145–13151
83. Zhou P, Chen B, Davis FA (2004) Tetrahedron 60:8003–8030

Index

Symbols
(3-aminopropyl) triethoxysilane (3-APS), 64, 65, 68, 69, 71
α-chymotrypsin, 29, 30, 31, 35

A
Abiotic, 3
Active site, 11, 36, 39, 42, 63
  considerations, 33, 35
Adsorption, 42, 48, 49, 50, 63
Agarose, 46, 47, 49, 54
Agricultural, 6
Aldol reaction, 19, 25
Aldose, 20
Alkoxysilane, 29, 31, 36
Allyltrimethoxysilane (ATMS), 31
Amino acids, 19, 35, 53, 103, 105, 106, 120
Angiotensin converting enzyme, 11
Anomeric, 22, 24
Antibiotic, 13
Antiseptic, 13
Arrhenius factor, 81

B
Bacteria, 3, 45
Banana, 2
Binding pocket, 33, 88
Biocatalyst, 39, 41, 63, 74, 86
Biocompatibility, 7, 13, 53, 73
Biodegradability, 13
Bioisostere, 11
Biological, 2, 6, 39, 109
Biomedical, 6, 12
Biosilica
  synthesis, 27, 28, 29
Biosilicates, 94, 101
Biotechnology, 6
Biotic, 3
Biotransformation, 63
Bond
  energy, 73
  length, 2
  strength, 2
Bovine serum albumin, 29, 51
Bromelain, 29
Brook rearrangement, 105, 106

C
Camptotheca acuminata, 12
Cancer, 10, 12, 108, 109
Candida antarctica, 64, 65, 73
Candida antarctica lipase B (CALB), 71
Candida rugosa, 29
Carbohydrate, 3, 21, 23, 53
Carbonic anhydrase, 29
Carboxypeptidase, 29, 33
Cathepsin, 4
Celite, 49
Chemotherapy, 12, 13
Chlorotrimethylsilane (CTMS), 64
Contact lens, 13
Contraceptive, 13
Controlled release, 13, 41, 93
Cross-linked enzyme aggregate (CLEA), 50
Crosslinked sol-gel materials, 101
Crosslinking, 7, 64, 65, 67, 94, 96, 101
Cucumber, 2
Cyclooxygenase (COX), 10
Cylindrotheca fusiformis, 55
Cysteine protease, 29
Cytochrome P450, 2

D
Detergent, 45, 63
Diatom, 3, 4, 6, 21, 29, 55
Diffusion coefficient, 14

P. M. Zelisko (ed.), *Bio-Inspired Silicon-Based Materials,* Advances in Silicon Science 5,
DOI 10.1007/978-94-017-9439-8

Diglycerylsilane (DGS), 52
Dopamine, 11, 104
Drought, 3
Drug, 11, 13

E
Elastomer, 14
Encapsulation, 42, 51
Entrapment, 42, 51, 52, 63
Enzyme, 2, 11, 29, 30, 33, 35, 39, 40, 42, 47, 49, 51, 53, 74, 111
Epichlorohydrin, 96
Epidermal cells, 2
Epoxy-amine crosslinking, 93, 95
Estring, 13
Ethyltrimethoxysilane (ETMS), 31

F
Food, 7, 45, 63
Formaldehyde, 19, 20, 24
Formose reaction, 19, 20, 23, 24
Fungi, 3
Furanose, 21, 22, 23

G
Gene expression, 3
Germanium, 12
Glass transition temperature, 14, 73, 81
Glutaraldehyde, 47, 50, 51, 65, 67, 71
Glyceraldehyde, 20, 24, 25
Glycidoxypropyltrimethoxysilane (GPTMS), 47, 92
Glycitol, 21
Glycoladehyde, 19, 25
Glyconic acid, 21
Gonadotropin-releasing hormone (GnRH), 12, 108

H
Haloperidol, 11, 104
Herbicide, 6
Herbivore, 3
Heteroatom, 1
HIV protease, 11, 109, 110
  inhibitor, 112
Horsetail, 2
Human serum albumin, 29
Hydrolysis, 4, 29, 31, 85, 111, 115

I
Immobilisation, 40, 42, 45, 47, 48, 49, 51, 54, 55, 58
  techniques, 44
Implants, 13
Indomethacin, 10
Inhibitor, 29, 108, 109, 116, 121
Insects, 3
Interpenetrating network (IPN), 94, 95

K
Karenitecin, 12
Kinetics
  elongation, 74, 77

L
Lipase, 29, 44, 45, 46, 47, 49, 55, 67
Lipophilic, 10
Long chain polyamine (LCP), 29
Lumen cells, 2
Lysozyme, 29

M
Marine organisms, 3, 4, 27
Medicine, 7, 10
Metabolism, 104
Methyltrimethoxysilane (MTMS), 31, 32, 92
Michalis-Menton, 45
Microorganisms, 3
Mildew, 3

N
Nanostructure, 21
N-isopropylacrylamide, 54
Nitzshia curvilineata, 4
Non-steroidal anti-inflammatory drug (NSAIDs), 10
Norplant, 13
Novozym-435 (N435), 46, 50, 65, 67, 69, 70, 71, 74
Nuclear magnetic resonance (NMR), 67, 75

O
Oligosaccharide, 23
Organically modified silicon materials (ORMOSILS), 91
Organosilicate, 95, 100, 101
Organosilicone, 6

P
Pancreatic carcinoma, 10
Papain, 4, 29
Parkinson's disease, 11
Pentacoordinate, 22
Pepsin, 29, 30, 33, 35
Peptidomimetic, 121
Perfume, 63

Phenyltrimethoxysilane (PTMS), 30
Phthalocyanine 4 (Pc 4), 10
Physiology, 7, 10
Phytolith, 2
Plants, 2, 3, 45
Polyallylamine (PAH), 55, 65
Polyamidoamine (PAMAM), 55
Polycarbonate, 63
Polycationic pepties *See* Silaffins, 4
Polyethylenimine (PEI), 48, 65
Poly-L-lysine (PLL), 55
Polymerization, 64, 74, 75, 78, 109
Polypeptide, 105
Polysiloxane *See* Silicone, 109
Polystyrene, 48, 49
Prebiotic, 23, 24
Predation, 4
Propyltrimethoxysilane (PTMS), 64
Protease, 4, 103, 116
Protodesilylation, 111
Pyranose, 21, 22

R
Radiolarian, 21
Reinforcement, 3
Resin, 42, 43, 46, 70, 74
Ribonucleic acid (RNA), 2
Rice, 2
Ring-opening polymerization (ROP), 64

S
Salinity, 3
Schizophrenia, 11
Science fiction, 1
Secondary metabolites, 3
Serine hydrolase, 29
Serum, 10, 51
Silacidins, 27, 28, 29
Silaffin, 4, 55
Silanediol, 103, 109, 110, 111, 114, 118
    synthesis, 111, 112
Silanization, 64
Silanol, 29, 76
Silica, 2, 4, 29, 42, 58, 66
Silica protein\t See Silicatein, 4
Silicate, 21, 22, 25, 97
Silicatein, 29, 35
Silicic acid, 2, 3, 4, 27, 29
Silicon, 1, 2, 3, 6, 29, 76, 107, 109
Silicon-based life, 1, 2
Silicone, 6, 13, 73, 75
Silwet\t See Superwetter, 6
Skeletal system, 3
Sol-gel, 52, 53, 101
Sorghum, 2
Spicules, 4, 27
Sponge, 2, 4, 6, 21, 29
Starch, 94, 95, 96, 101
Star Trek, 1
Stomata, 6
Sugar, 19, 21, 22, 23, 96
Superwetter, 6
Surfactant, 6, 51

T
Terrestrial organisms, 2
Tethya aurantia, 4
Tetraethoxysilane (TEOS), 4, 29
Thermal decomposition, 97, 101
Thermal gravimetric analysis (TGA), 97
Thermolysin, 110
    inhibitor, 113, 114
Thorns, 3
Trypsin, 6, 29, 30, 31, 32, 33, 35

V
Vaginal ring, 13, 15
Vertebrate, 7

W
Weinbaum, Stanley, 1
Wells, H.G., 1
Wheat, 2

X
X-ray, 21, 23

Z
Zeolite, 2, 46

Zeitfracht Medien GmbH
Ferdinand-Jühlke-Straße 7
99095 Erfurt, Deutschland
produktsicherheit@kolibri360.de